Jürgen Schulte

DAS JAGDLEXIKON

Ein jagdliches Wörterbuch einschließlich Jägersprache

DEUTSCH - ENGLISCH
ENGLISCH - DEUTSCH

Wendebroschüre

Impressum

ISBN 978-3-7888-2073-2

1. Auflage 2023
Printed in Germany

Erschienen im Ressort
Jagd-Kultur
im Auftrag des Verlages
J. Neumann-Neudamm

c/o NJN Media AG
Unter dem Schöneberg 1
D-34212 Melsungen

info@neumann-neudamm.de
www.neumann-neudamm.de

Bildnachweis: Lizensfreie Shutterstock-Fotos.

VORWORT

Seit 1759, dem Jahr, in dem Johann August Großkopff das erste „Neu und wohl eingerichtete Forst-/ Jagd- und Weidewerks-Lexikon“ veröffentlichte, hat es viele deutschsprachige Lexika gegeben. Alle stellen die umfangreiche jagdliche Fachsprache vor.

Die Internationalisierung jagdlicher Betätigung veranlasst viele im Ausland jagende Jäger, sich mit ihren ausländischen Betreuern zu verständigen. Hierzu soll das vorliegende Fachwörterbuch beitragen.

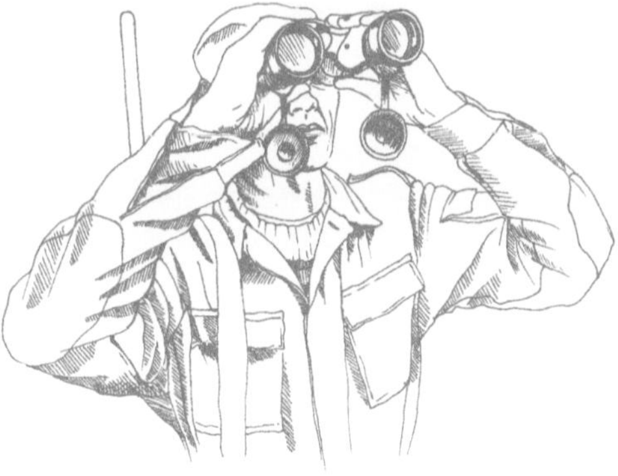

DEUTSCH - ENGLISCH

A

Aalstreifen, Aalstrich	eel back, eel stripe, line along back of animals, dorsal stripe
Aasjäger, Schießer	pot hunter
Aaskrähe	carrion crow
abbalgen, abstreifen, streifen	to skin
abbalzen	to end courtship | to end mating
abbaumen	leaving the tree (bird) | leaving the high seat (hunter)
abbeißen: Vertreiben ausgewachsener Jungtiere durch Raubwildeltern	to bite young adult game by predaceous parents
abblasen	to blow off | to announce the end of hunt by signal | to call off
abbrechen	to break off | to snap off
abbrunften	to cease rutting | to end rutting
abdanken	thanks to the hunters and beaters (drivers)
abdocken	to unroll the leash
abdrücken	to pull the trigger
Abendansitz	still hunt at dusk
Abendbalz	courtship in the evening | mating the evening | display in the evening
Abendstrich	passage in the evening
abfährten, abspüren, spüren	to track | to follow the trail of animals
abfallen. 1) nach dem Beschlag vom weiblichen Stück ablassen (auch absitzen). 2) s. abbaumen (Verlassen eines Baumes durch Waldhühner)	1) to descend from hind after covering (stag). 2) to fly off (feathered game)

abfangen, abnicken, Fang geben	to kill a wounded hoofed game by stabbing with a hanger, to stab
abfedern (Tierquälerei)	to kill a wounded feathered game by using a flag feather, pushed into the back of the head
abführen, führen, abrichten	to train, to train the dog for hunting, to break in
abglasen, ableuchten	to scan
abhalsen, schnallen	to uncouple, to unleash, to let loose
abkämpfen (Stangen oder Geweih im Kampf abbrechen)	the loss of a branch of antlers by a stag while fighting against a rival
abkochen	to boil, to scald
abkommen	1) to loose condition (game). 2) to aim by shooting on the mark
ablegen	to lie down (dog)
ableuchten, abglasen	to scan
abnehmen, abtragen	to take off the hunting dog from the track
abnicken, abfangen, Fang geben	to kill a wounded hoofed game by stabbing with a hanger, to stab
abnorm	abnormal, irregular
abpassen	snap shooting
abreiten, abstreichen	to fly off
abrichten, abführen, führen	to train, to break in, to train the dog for hunting
abrufen	to call off a hunter or a dog
abschärfen	to strip the skin, to skin
abschießen	to shoot the game in a district
abschlagen	1. to bay the dogs by stags or wild boars. 2. to beat off a rival by stag or wild boar. 3. to punch (hare, rabit)

Abschuss, Gesamtabschuss	the whole bag of hooofed game, shooting
Abschusshirsch	cull stag, a stag with deformed or bad antlers
Abschussplan	shooting plan
abschwarten	to skin (for wild boar)
Absehen	aiming point
absetzen	to break off shooting
absprengen	to cut off (forced separation red deer)
abspringen, fliehen, flüchten, ausbrechen	leap (to) off, to escape, to run away, to flee
abspüren, abfährten, spüren	to track, to follow the trail of animals
abstellen, umstellen	to post the hunters and beaters around the hunting district, to surround
abstreichen (abreiten) Federwild	to fly off (feathered game)
abstreifen, abbalgen, streifen	to skin
absuchen, Suche	tracking, dogging, to search
abtragen, abnehmen	to take off the hunting dog from the track
abwerfen	to shed the antlers, to cast
Abwurf	shedding of antlers
Abwurfstange	shed antler
abwürgen	to kill small game after pulling down by hunting dog
Abzug	trigger
Abzugsbügel	trigger guard
Abzugseisen	steel trap, trap
Abzugssicherung	trigger safety
Achter, Achtender	eight points stag
Adler	eagle
Adlerhorst	eagle´s eyrie
Afterklauen, Geäfter, Oberrücken	hind claws, dewclaw, ergot

Ahnentafel	genealogical table, ancestral register
Allesfresser	omnivore
Alpendohle	Alpine chough
Alpenkrähe	chough
Alpenschneehuhn	ptarmigan
Alpensteinbock	Alpine ibex
alt	old, aged
Altersbestimmung	age determination
Altersklasse	age class
Alterskleid	adult plumage of wild fowl
Alttier, Rottier, Tier	adult animal, female adult animal, female red deer, hind
anbacken, anschlagen	to aim
anblasen	to blow on, to call on
anbringen, antragen (Schuss)	to shoot at game
anbrüchig, faulend	to rot
anfahren	to approach game in a boat or horse-carrage
anfallen (Annehmen der Fährte durch den Schweißhund)	to follow track (for the dog)
anfliehen	to fly against (trees, hunters, fences), to bump
Angstgeschrei	alarm sound
ankirren, anludern, ködern	to put food for the game, to lure with a bait (carcass), to bait, to decoy, to entice, to lure
anlaufen	to run towards the hunter
annehmen (Futter)	to accept feeding
annehmen, angreifen	to attack
anpirschen	to stalk game
anrufen	to call the game for a suddenly stop
anschießen (Gewehr)	to test the rifle or shotgun
anschießen, anschweißen, krankschießen	to wound a game
Anschlag	gun mounting, firing position, aiming position

anschlagen, anbacken	to aim
anschlagen, bellen, Laut geben, verbellen	to bark, to give tongue
anschneiden	the dog or the vermin devours a shot game, to eat
Anschuss	place where game stands while being shot at
anschweißen, anschießen, krankschießen	to wound
Ansitz, Anstand	sitting, stand hunting
ansitzen, ansetzen	to sit up for shooting
Ansitzleiter	ladder
ansprechen	to detect the details (age, gender, track and so on) of game, to identify
anspringen	to stalk a calling capercaillie by bounds
Anstand, Ansitz	stand hunting,sitting
anstellen	to assign the place for hunters and beaters
anstreichen	1) to support the riffle against a tree. 2) to approach (feathered game)
Antilope	antelope
antragen, anbringen	to shoot at game
antreiben	beginning of drive
anziehen	to use his nose for approaching the game (hunting dog)
Appell, Gehorsam	obedience of the hunting dog
apportieren, bringen	to retrieve, to bring, to fetch
Apportierhund	retriever
Arbeit	work
Armschwinge	secondary feather
Artenschutz	protection, conservation (preservation) of species
äsen	to browse, to graze
Äser, Geäse	mouth
Äsung, Futter	food, feed, fodder

Äsungsfläche	feeding ground
Attrappe	decoy
atzen	to feed up, to rear
Atzung	food for the bird of prey
Auerhahn, Urhahn	capercaillie
Auerochse	wildox, aurox, urus
Auerwild	wood (great) grouse, capercaillie
aufbaumen	to perch
aufbrechen, Ausfahren von Waldhühnern	to crack, to gut (for capercaillie), to disembowel, to dehisce, to eviscerate
Aufbruch zur Jagd	departure for the hunt
Aufbruch, Jägerrecht	bowels, entrails
Aufgang der Jagd (Ende der Schonzeit)	beginning of the open season
Aufgeben der Fährte durch den Hund	to leave the track (hunting dog)
aufhaben	to wear (a stag wears good or bad antlers)
aufmüden	to rouse wounded game
aufnehmen 1) Hund nimmt Fährte auf. 2) Trächtigwerden von Raubwildfähe und Hündin. 3) Sauen nehmen Fraß auf	1) to follow a trail of game. 2) to be pregnant (bitch). 3) to take food (wild boar)
aufschärfen, schärfen	cut (to) up or to gut the skin of furred game or the plumage of winged game, to blade up
aufsetzen: 1) Trophäe auf Holzbrett montieren. 2) Geweih schieben	1) to mount the trophies on a bard, to attach. 2) the beginning of pushing the antlers
aufsitzen, beschlagen	to cover or to get on
aufstehen	to get up
Auftreffenergie	energy of impact
aufwerfen	to throw up (for all hoofed game, excluding wild boar)

Auge, Licht, Seher	eye
äugen	to look at
Augsprosse	brow tine
ausbrechen, fliehen, flüchten	to escape, to flee, to run away
ausbrüten, ausbringen	to hatch, to brood, to incubate
Ausfahren. 1) Verlassen des Baues, der Sasse, des Lagers. 2) Aufbrechen oder Ausfahren, Ausnehmen von Waldhühnern, (Auerwild)	1) to leave the burrow (fox and badger), the lair (hare and wild boar). 2) to disembowel, to dehisce, to eviscerate, to gut (for capercaillie)
ausfallen, schlüpfen	to hatch out, to moult, to come out
Ausfuhrgenehmigung	export licence
ausgehen	to follow the track of marten in the snow
ausgraben	to dig up
aushaken, ausnehmen, ausziehen (aushaken nicht mehr gebräuchlich)	to draw out (uncommon), to pull out, to extract, to remove
Auslage	span, spread
auslaufen (Nestflüchter verlassen das Nest)	to leave the nest by precocial
Auslösen der Grandeln beim Rotwild	to take out the upper canine tooth (trophy) of red deer
ausmachen, bestätigen	to find out, to ascertain, to locate
ausnehmen (aufbrechen, Ausfahren von Waldhühnern)	to gut, to disembowel, to dehisce, to eviscerate
ausneuen, ausgehen	to follow the track of marten in the snow
Ausrüster	outfitter
Ausrüstung	equipment
ausschießen	to clear out by shooting all game in a shooting-ground
Ausschuss	exit hole
ausschweißen, schweißen	to bleed
Außenballistik	exterior ballistics
aussetzen, auswildern	to release

aussteigen	to get out of water (for otter and beaver)
austreten, auswechseln, treten	to leave or to change the home district (wood) for searching feed (for hoofed game)
Auswerfer, Ejektor	ejector
auswildern, einbürgern	to introduce
ausziehen, aushaken	to pull out, to extract, to remove
Auszieher	extractor
Axishirsch	spotted deer, chital

B

Bache	wild sow, cheek
Backe	cheekpiece
Backenzahn, Molar	molar, molar tooth
Bakterie	bacterium
Balg	skin, coat, fur, pelt
Ballen	ball
Ballistik	ballistics
Balz	courtship, mating, display
Balzarie, Balzgesang, Balzlaut, Gesetzl	love song, courtship call, courtship / mating song
Balzbaum	tree of courtship or of mating
balzen	to court, to pair, to copulate, to display
Balzfedern, Balzstifte, Zehenstifte	teeth of comb
Balzflug	mating flight
Balzgesang, Balzarie, Balzlaut, Gesetzl	love song, courtship call, courtship / mating song
Balzjagd	shooting in the mating season
Balzkleid	courtship plumage

Balzkragen	throat feathers
Balzlaut, Balzarie, Balzgesang, Gesetzl	love song, courtship call, courtship / mating song
Balzlosung, Balzpech	dung of mating
Balzplatz	display area, dancing ground
Balzrose, Flamme, Rose	red spot above the eys of black grouse and capercaillie, eye or red wattle
Balzspiel, Balzverhalten	courtship behaviour
Balzstellung	courtship (display) posture
Balzstifte, Balzfedern, Zehenstifte	teeth of comb
Balztanz	mating (courtship) dance
Balzverhalten, Balzspiel	courtship behaviour
Bär	bear
bären	to cover (bear and marmot)
Bart	chamois beard, beard
Bartgams	chamois buck with long hairs on the back
Bartgeier	bearded vulture
Bärzeit	time for pairing (bear)
Basküle	hinged frame
Bast, Gefege	velvet
Bastgeweih	velvet covered antlers
Bätze, Betze, Fähe, Füchsin	female of canids (fox), bitch fox, vixen, fuchsin
Bau	burrow
Bauarbeit	earth work
Bauch, Magen	belly, stomach
Bauhund, Erdhund	earth dog
Baujagd, Erdjagd	earth hunting, bolting
baulaut	to give tongue in the burrow without contacting to the game
Baumbrüter	hole breeder
Baumfalke, Lerchenfalke, Schwalbenfalke	hobby

Baummarder, Tannenmarder	pine marten
Bayerischer Gebirgsschweiß-hund	Bavarian bloodhound
Beagle	beagle
befahren (bewohnter Bau)	to live in the burrow, occupied
begehren	the call of rutting hinds
Behang	ears of dog
Beihirsch	second stag
beißen	to bite
Beitritt	overreaching
beizen	to hunt with a falcon
Beizjagd, Beize	hawking
Beizvogel	hunting bird
bejagen	to hunt
Bekassine, Himmelsziege, Schnepfe	snipe
bellen, Laut geben des Jagdhundes, anschlagen, verbellen, auch Lautäußerung des Fuchses während der Ranz	give (to) tongue (hunting dog), to bark, to yelp, also by foxes during the rutting season
Benehmen, Verhalten	behaviour
beringen	to ring
Berufsjäger	professional hunter
Besatz (gilt nur für Niederwild), siehe auch Wildbestand	stock of low game
Besatzdichte	stocking rate
beschießen	to shoot
Beschlag	act of covering, copulation
beschlagen	to cover, to get on
besetzen	to post the guns
Bestand, Wildbestand	stock of game
bestätigen, ausmachen	to find out, to ascertain, to locate
Bett	bed
Betze, Bätze, Fähe, Füchsin	female of canids (fox), bitch fox, vixen, fuchsin
Beute, Strecke, Jagdstrecke	bag, prey, booty, capture

bewinden	to smell
Biber	beaver
Biberburg	castle of beaver
Biberkelle, Kelle	beaver´s tail, beaver slice
Biberratte, Nutria, Sumpfbiber	coypu, nutria
binden, hängen (Hund, Fuchs u. Wolf nach Verpaarung)	to hang together (dog, fox, wolf after copulation)
Birkhahn, Spielhahn	blackcock
Birkhenne	greyhen
Birkwild	black grouse
Birsch, Pirsch, Pürsch	stalking
Bisam	musk
Bison	bison
blädern (Gamswild)	to bleat (chamois)
blanke (kalte) Waffen	cold steel, cold weapons (hunting knife)
Blase	bubble, blister, bladder, vesicle
blasen 1) Schwarzwild. 2) Jagdhorn	1) to blow (wild boar), to puff. 2) to blow (hunting horn)
Blässgans	white-fronted goose
Blässhuhn	coot
Blatt 1) Schulterblatt, 2) Schild	1) shoulder blade (wild boar), 2) shield (capercaillie, hazel cock, cock pheasant, partridge)
blatten, fiepen	to call, to bleat, to whistle
Blatter, Rehlocker	caller, roe deer call
Blättermagen	psalterium, third stomach of ruminants, omasum
Blattjagd	roe deer calling
Blattschuss	shoulder shot, chest hit
Blattzeit	rutting season of roe
bleichen	to bleach
blenden	to dazzle
Blendling (Kreuzung zwischen Haus- und Wildkatze)	cross breed (cross-breeding between cat and wild cat)

blind (Gamsbart ohne Reif)	blind
Blume (Schwanz des Hasen, Luntenspitze)	tail of hare, tag for the end of fox tail
Bock (Rehbock)	roe buck
Bockbüchsflinte	combination gun
Bockdoppelbüchse	over-under rifle
Bockdoppelflinte	over-under shotgun
Bockkitz	male kid
Bodenbalz	to call on the ground during the coupling season (capercaillie)
Böhmische Streife	walking up
Borsten, (Federn f. Schwarzwild), Saubart	bristles (back of wild boar)
Brachvogel, Großer	curlew
Bracke, Wachtelhund, Spaniel	spaniel
brackieren	hunting with spaniels
Brandadern	crural veins
Brandente	shelduck
Brandfuchs	black coloured red fox
Branke, Brante, Pranke, Prante	claw, clutch, pad, paw
Brauchtum	custom
Braunbär	brown bear
Braunschimmel (Brauntiger) häufig bei Deutsch Drahthaar und Deutsch Kurzhaar	grey/white coloured hunting dog with brown patches
brav	good
brechen	to root by searching for food (wild boar)
breit (Profilstellung des Wildes)	broadside on
bringen, apportieren	to retrieve, to bring, to fetch
Brocker (Schnabel Auerwild)	bill
Bruch	branch, broken branch
Bruck (Unterschwanzdecken des Greifvogels)	under tailcoverts
Brunft, Brunst	rutting season, heat period

brunften	to be on heat
Brunftfeige, Duftdrüse, Nelke, Viole	scent gland
Brunftfleck	rutting mark, rutting spot
Brunftgeruch	rutting smell
Brunfthirsch	roaring stag
brunftig	to be on heat
Brunftkugel, Kurzwildbret	coddes, testicle
Brunftkuhle (Damhirsch)	depression (hollow) on the ground during the heat of the fallow buck
Brunftmähne, Mähne, Kragen	rutting-mane, mane
Brunftplatz	rutting place
Brunftrute, Feuchtglied, Fruchtglied (Hund, Haarraubwild), Rute (Hund, Haarraubwild)	penis, pizzle (predaceous animals)
Brust	chest, breast
brüten, ausbringen	to hatch, to brood, to incubate
Brutfleck	incubation patch
Brutplatz	nesting place, breeding place
Brutzeit	breeding season, brooding time
Büchse	rifle
Büchsenlauf	rifled barrel
Büchsenlicht	shooting light
Büchsenmacher	gunmaker
Büchsenpatrone	rifle cartridge
Büchsenschuss	shot of rifle
Büffel	buffalo
Burgstall	hillock
Bürzel, Pürzel, Federlein (Schwarzwild)	tail (wild boar)
buschieren	dogging
Bussard, Mäusebussard	buzzard

C

Chokebohrung, Würgebohrung	choke boring
coupieren, kupieren	to cut a tail, to dock

D

Dachs	badger
Dachsbau	badger´s set
Dachsfett, Schmalz, Weißes	fat of a badger
Dachshund, Dackel, Teckel	dachshund
Dächsin	female badger
Dachslosung	fuants
Dachsschwarte	badger skin, hide
Dackel, Dachshund, Teckel	dachshund
Dallschaf	dall sheep
Damhirsch, Damschaufler	fallow stag, fallow buck
Dämmerung	dusk
Damtier	fallow doe
Damwild	fallow deer
Darm	intestine, bowels, gut
Dasselfliege	warble fly
Dauergebiss	permanent dentition
Decke (Schalenwild, außer Schwarzwild – Schwarte)	skin, hide, coat, leather
decken: 1) Saupacker greifen Schwarzwild; 2) Zusammenhalten der Schrote bei nicht streuender Flinte	1) to attack a wild boar (hunting hounds), 2) spread cone for a shotgun
Deckhaar	straight hair, kemp
Deckung	cover, camouflage, hide
Deutsch Drahthaar	German wirehaired pointer
Deutsch Kurzhaar	Gernan short-haired pointer
Deutsch Langhaar	German long-haired pointer

Deutsch Stichelhaar	German rough-haired pointer
Deutsche Bracke	German beagle
Deutscher Jagdterrier	German terrier
deutscher Stecher, Stecher	double set trigger
Deutscher Wachtelhund	German springer
dickgehen, trächtig, hochbeschlagen, innehaben	to be pregnant
Doppelabzug	double trigger
Doppelbüchse	double rifle
Doppelflinte	double shotgun, double-barrelled gun
Doppelkrone	double crown
doppeln	double
Doppelschnepfe	great snipe
Dorn, Schauflerhaken	the tine on the ground of the shovel antler; back antler
Down! (nieder!)	down!
Drahle	varvel
Drall	rifling
Dreiläufer (¾ ausgewachsener Hase, halbwüchsig)	leveret (3 month old)
Dreiläufer, Drilling	three-barrelled gun
Drilling, Dreiläufer	three-barrelled gun
Drossel, Luftröhre	trachea, windpipe
Drosselknopf, Kehlkopf, Stimmknoten	larynx
Drücken: 1) Verstecken des Wildes, 2) Wild aus dem Treiben drücken	1) to hide, 2) to rouse
Drückjagd	rousing game
Drückjagdbock	shooting seat
Drückjagdvisier	driven game sights
Druckpunktabzug	double pull trigger
Ducker	duiker
Duftdrüse, Nelke, Viole, Brunftfeige	scent gland

durchgehen, durchtreiben, abtreiben, treiben	to drive small game
Durchgehschütze	beater shooter
durchrinnen, rinnen	to swim across
durchziehen	to traverse

E

Eckzahn, Fangzahn	canine tooth
Edelmarder	pine marten
Edelwild, Rotwild	red deer
Eichelhäher, Häher	jay
Eiderente	eider
Eigenjagd	private hunting ground
Einabzug	single trigger
einfahren (Aufsuchen des Baus)	to go to earth or to ground
einfallen (Einfahren des Dachses)	to go to earth, to go to ground
einfallen (Vogel lässt sich nieder)	to perch, to settle
Einfuhrgenehmigung	import licence
eingehen	to die
Eingriff (Schalenabdruck)	slot
Einhakmontage	hook in mount
einheimisch	indigenous
einläufige Flinte	single shot
einschieben	to lie down
einschießen	to zero
einschliefen	to slip in
Einschuss	shooting spot, entrance hole
einschwingen (Auerwild lässt sich nieder)	to alight on a tree
Einstand (Schalenwild)	lair (hoofed game), harbour
einstechen, stechen	to prick, to set the trigger

Einstecklauf	insert barrel
Einzellader	shotgun
Eisbär	polar bear
Eisen, Fangeisen, Abzugseisen	trap, steeltrap
Eisente	long-tailed duck
Eisfuchs, Polarfuchs	arctic fox
Eissprosse	bay tine
Ejektor, Auswerfer	ejector
Elch	elk, moose
Elch-Alttier	elk-cow
Elchhund	elk hound
Elefant	elephant
Elefant (Afrikanischer)	African elephant
Elefant (Indischer)	Indian elephant
Elenantilope	eland
Elfenbein	ivory
Elster	magpie
Endballistik, Zielballistik	terminal ballistics
Endbinde, Stoßendbinde	sub terminal band
Ende, Geweihzacken, Sprosse	branch, prong of antler, tine, point
englischer Schaft	english stock
Englischer Setter, Lavarack	English setter
Ente	duck
Entenstrich	flighting of ducks
entladen	to unload
entsichern	to unlock
entspannen	to half-cock
entstechen	to uncock the trigger
eräugen, wahrnehmen	to see
Erdhund, Bauhund	collective name for dachshounds and terriers, earth dog
Erdjagd, Baujagd	earth hunting, bolting
erlegen, strecken, erbeuten	to kill, to bag
Erleger, Jäger, Schütze, Weidmann	hunter, marksman, markswoman, shooter, gunner
Erntehirsch	trophy stag

Erpel	drake
Erpellocke	curl of drake
Erste Hilfe	first aid
Eule	owl

F

Fächer	fan, tail
Fadenkreuz	cross wire
Fähe, Bätze, Betze, Füchsin	female of canids (fox), bitch fox, vixen, fuchsin
Fahne, Lunte, Rute, Standarte	breech, brush, drag, tail
Fährte, Spur, Wildspur	trail, track, spoor
fährten	to track
Fährtenkunde	knowledge of tracks and trails
Falke	falcon
Falkenbeize	hawking, falconry
Falkenhaube	hood
Falkner	falconer
Falknerei	hawking
Falknerhandschuh	falconers glove
Falknertasche	hawking bag
Falle	trap, snare
Fallenjagd	trapping
Fallensteller	trapper
Fallwild, Luder	carrion
Fang (fangen)	catching of game
Fang (für Greifvögel, Gewaff)	claw, talon
Fang geben, abfangen, abnicken	to kill a wounded hoofed game by stabbing with a hanger, to stab
Fang, Maul	muzzle, mouth
Fangeisen, Eisen, Abzugseisen	trap, steeltrap
fangen	to catch, to capture
fängisch	ready for catching

Fangjagd	trapping
Fangschuss	coup de graze, finishing shot
Fangzahn, Eckzahn	canine tooth
Farbe	the blood of furred vermin
färben (Haarwechsel und Schweißen der läufigen Hündin aus der Schnalle)	moult, moulting if a female dog on heat bleeds from vagina
Fasan	pheasant
Fasanenhahn	cock pheasant
Fasanenhenne	hen pheasant
Fasanerie	pheasant house
Federbalg	feather follicle
Federkleid	plumage
Federlein, Bürzel, Pürzel	tail (wild boar)
Federn (Rückenhaare oder Saubart des Schwarzwildes), Borsten	bristle (back of wild boar)
federn, krellen	to stun, to hit the spine
Federohren, Hörner	plumicorn, feathered ear, ear tuft
Federspiel	lure
Federwechsel, Mauser, Raue	moult, moulting
Federwild	feathered game
fegen	to rub off, to fray
Fegeschaden	fraying damage
Fegestelle	velvet mark
fehlen	to miss
Fehlschuss	miss
feist	obese, fat, stout
Feist (Unschlitt)	grease, fat
Feisthirsch	stag in grease
Feistzeit	venison season
Feld: 1) gezogener Lauf, 2) Altersangabe des Jagdhundes	1) land (shotgun), 2) field (age of hunting dog)
Feldhase, Hase	hare
Feldjagd	field hunting

Fernrohr, Fernglas, Jagdglas, Prismenglas	binocular, field glass, telescope
Fessel, Langfessel	leash
Feuchtblatt	vagina, vulva of doe (hind)
feuchten, nässen	to piss, to urinate
Feuchtglied, Brunftrute, Fruchtglied (Hund, Haarraubwild), Rute (Hund, Haarraubwild)	penis, pizzle (dog, predaceous animals)
Feuer (Wild liegt im Feuer)	game falls in the fire
fiepen, blatten	to call, to bleat, to whistle
Finder, Saufinder, Packer	boarhound, trailer, tracker
Fischadler	fish-hawk, osprey
Fischotter	otter
Fischreiher, Graureiher	grey heron
Flamme, Rose des Auer- und Birkhahnes, Balzrose	red spot above eyes of capercaillie and blackcock, eye wattle, red wattle
Flanke	flank
Fleckschuss	point blank shot
fliehen, flüchten, ausbrechen	to flee, to escape, to run away
Flinte	shotgun
Flintenabzug	single stage trigger
Flintenlauf	shotgun barrel
Flintenlaufgeschoss	slug
Flintenschütze	shotgun shooter
Flucht	flight, escape
Fluchtbau, Notbau	burrow for taking flight, emergency burrow
flüchten, fliehen, ausbrechen	to escape, to flee, to run away
Fluchtfährte	trail of a fleeing game
flüchtig	fleeing, volatile, on the run
Flug	a group of feathered game, flight
Flugbahn	trajectory
Flügel, Schwinge	wing, pinion
flügeln	to wing

flügge	fledged, ready to fly
Flugrichtung	path of the target
flugunfähig	flightless
Flugwild	winged game
Flugwildjagd	wingshooting
Folge (Richtung)	direction
folgen	to trail, to follow
forkeln	to spear, to gore, to butt
Foxhound	foxhound
Foxterrier	fox-terrier
Fraß	feed, feeding, food material
Fraßschaden	damage caused by pests, pest-damage
freihändig	free-handed
fressen	to feed, to devour
Frettchen	ferret
frettieren	to ferret
Frettierjagd	ferreting
Frettiernetz, Kaninchenhaube, Netz	purse net
frisch (frische Fährte)	fresh trail
frischen	to farrow, to litter
Frischling	young wild boar, gruntling, piglet
Frosch (Frischling bis Alter von 4 Wochen)	piglet
Frostgeweih	chilled antlers
Fruchtglied (Hund, Haarraubwild), Brunftrute, Feuchtglied, Rute (Hund, Haarraubwild)	penis, pizzle (predaceous animals)
Fuchs, Rotfuchs	fox | red fox
Fuchsbalg	fox skin
Fuchsbandwurm	fox tapeworm
Fuchsbau	fox hole
Füchsin, Bätze, Betze, Fähe	female of fox, vixen, fuchsin
Fuchslosung	billeting, billets

Fuchsreizen, Reizjagd	fox calling, decoying a fox
Fuchswelpe, Jungfuchs	fox cub
führen (Gewehr führen)	to use a gun
führen (Jagdgast führen)	to lead a shooting-guest
führen (Jungwild durch Muttertier führen)	to go with parents-game
führen, abführen	to train
Fuß	foot
Futter, Äsung	feed, food, fodder
Futteral	case
füttern	to feed
Futterstelle	feeding place
Fütterung, Wildfütterung	game feeding

G

Gabelbock	fork buck
Gabelbock (Antilocapra americana)	pronghorn
Gabelgeweih	fork antlers
Gabelhirsch, Gabler	fork stag, brocket
Gabelhuhn	cheeper
Gabelweihe (zool. Roter Milan)	kite
Galgen	bird sling
Gallenblase	gall bladder, cholecyst
Gams, Gamswild, auch Gemse	chamois
Gamsbart, Wachler	chamois beard
Gamsbock	chamois buck
Gamsgeiß	chamois doe
Gamskitz	chamois kid
Gamskrucken, Krickel, Krucken	chamois horn
Gamskugeln	hair balls
Gans	goose
Gänsegeier	griffon vulture

Gänsesäger	goosander
Ganter	gander
Gatter, Wildgatter	fence, holding paddock, railing
Gatterwild	carted deer
Gazelle	gazelle
Geäfter, Afterklauen, Oberrücken	hind claws, dew claw, ergot
Geäse, Äser	mouth
Gebet	prayer
Gebiss	dentition, denture, set of teeth
Gebräch	rooting place
Gebrauchshund	hunting dog
Gebrech	snout
Gefege, Bast	velvet (from animal horn or antler)
Gefieder	plumage
geflügelt	winged, broken winged
Gegenwind	headwind
Geheck	litter
Gehege	enclosed pasture, enclosure, paddock, preserve, reservation, warren
Gehirn	brain
Gehör	hearing, sense of hearing, ear
Gehörn (besser: Geweih)	antlers, horns
Gehorsam, Appell	obedience
Gehörschutz	hearing protection
Geier	vulture
Geiß, Ricke	female doe, goat, doe
Geißkitz, Rehkitz, Rickenkitz	doe kid, roe kid
gejagt	hunted, chased
Geläuf	tracks
Gelege (Brut)	nest (clutch) of eggs
gelt, unfruchtbar	barren
Gelttier	spayed animal, sterile hind
Gemse (s. Gams)	chamois
Genick	nucha, back of the neck, nape

Genickfang	rabbit punch
genossen machen	to reward
Gepard	cheetah
geperlt	pearled
gerade	even
Geraffel, Scharwild, (Gamsgeißen, deren Kitze und geringe Böcke)	herd of chamois, their kids and bucks
Geräusch	Internal organs of game: heart, liver, kidneys, lung, spleen
Gerfalke	gyr falcon
gering, schwach	of poor condition, weak, little, small
Gesäuge, Spinne	mammary glands of animals, udder
Gescheide	intestisnes, entrails
Geschleife: 1) Bauröhren von Fuchs und Dachs, 2) Schleppe	1) burrow, tunnel, 2) scent-bag (hunting dogs)
Geschmeiß, Losung	dung
Geschoss, Kugel	bullet
Geschröte (Hoden von Hunden und Haarraubwild)	testicle
Geschüh	jess
Gesetzl, Balzlaut, Balzarie, Balzgesang	love song, courtship call, courtship / mating song
Gesicht	face, sight, vision
Gesichtssinn	sense of vision
gespannt	cocked
Gesperre	covey, bevy
geständert	legged
Gestüber	droppings
gesund	If the hunter has missed a game, it is healthy
Gewaff, Hauer, Waffen (Schwarzwild)	tusk of wild boar, weapon
Gewehr	gun
Gewehre (untere Eckzähne des Keilers)	arm, file, tusk

Gewehrfutteral	gun cover
Gewehrkolben	buttstock
Gewehrlauf	barrel
Gewehrriemen	sling, carry strap
Gewehrschloss, Schloss	lock
Geweih, Grind, Haupt, Kopf	antlers, horn, head
Geweihbildung	development (formation, growth) of antlers
Geweihstange, Stange	beam
Geweihzacken, Ende, Sprosse	branch, prong of antler, point, tine
Gewichtl oder Gewicht (süddeutsch für Gehörn bzw. Geweih des Rehbocks)	horns
Gewölle	cast, pellet, vomits (of predatory birds)
gezogen	rifled
Gibbon	gibbon
Giraffe	giraffe
Gnu	gnu, wildebeest
Gockel	cock
Golden Retriever	Goldon retriever
Gordon Setter	Gordon setter
Gorilla	gorilla
Gössel	gosling
graben	to dig
Grandeln	canine tooth of red deer
Grannen	coarse hair
Graser, Lecker, Weidlöffel, Zunge	tongue
Graubär	grizzly bear
Graugans	greylag goose
Graureiher, Fischreiher	grey heron
Gravur	engraving
Greifvogel	predatory (raptorial) bird
Greifzehe	prehensile toe
Griff	grasp, grip, handle, hold
Griffon	griffon terrier

Grind, Geweih, Haupt, Kopf	antlers, head
Grisfuchs	grey fox
Grizzlybär, Graubär	grizzly bear
grob	imprecise, rough
grobes Schwein, Hauptschwein	big wild boar, old solitary
Großer Münsterländer	large Münsterländer
Großes Wiesel, Hermelin	stoat, ermine
Großtrappe	great bustard
Großwild, Hochwild	big game
Großwildjagd, Hochwildjagd	big game stalking, big game hunting
gründeln	to dive and search for food
Gruppenbalz	social courtship
Gummistiefel	rubber boots
günstigste Einschießentfernung (GEE)	optimum sighting in range
gut	good

H

Haar	fur, hair
Haarraubwild	furred predator
Haarwechsel	moult, moulting
Haarwild	fur, furred game
Habicht, Hühnerhabicht	goshawk
Habichtsadler	Bonelli´s eagle
Habichtskauz	Ural owl
Haderer	grinder, canine teeth in the upper jaw of wild boar, upper tusk
Hagel, Schrot	small shot
Häher, Eichelhäher	jay
Hahn	cock
Haken (Eckzähne der Bache und des Fuchses)	hook, canine tooth of the wild sow

Haken schlagen des flüchtigen Hasen	to dodge about
Haken, Grandeln	teeth, canine tooth of red deer
halber Wind (Seitenwind)	side wind
Halbhase	leveret (2 month)
Halbschaufler	half shovel head
halbwüchsig, Dreiläufer	leveret (3 month old)
Halsung	dog collar
halten (Fährte halten durch den Jagdhund)	to stick to the track
Haltepunkt	point of aim
Handfeuerwaffe	portable firearm
handscheu	hand shy
hängen, binden (Hund, Fuchs, Wolf nach Verpaarung)	to hang together
Hannoverscher Schweißhund	Hannover bloodhound
hart	tough, hard
Hase, Feldhase	hare
Haselhuhn	hazel hen
Haselwild	hazel grouse
Hasenjagd	battue, hare shooting
Hasenklage (Hasenquäke)	hare call, hare pipe
Hasenohr, Löffel	hare´s ear, spoon (fiction)
hasenrein	down to fur
Häsin, Setzhase	female hare
hassen (Stoßen von Greifen, Rabenvögeln und anderen Vögeln auf ihre Feinde)	to hate, to swoop
Hatz, Hetze	hunting with hounds, pack of hunting dogs, forced hunt
Haubentaucher	great crested grebe
hauen	to strike
Hauer, Gewaff, Waffen	tusk of wild boar, weapon
Haupt, Geweih, Grind, Kopf	antlers head
Hauptschlag (dritter Teil der Balzstrophe des Auerhahns)	the third verse of the courtship call
Hauptschwein	old solitary, singular, big wild boar

Hege, Wildhege, Wildschutz	game keeping, wildlife management, game protection
Hegebusch, Remise	grove, copse, small woods in the landscape for game shelter
hegen	to preserve, to protect
Heger, Jagdaufseher, Jagdhüter	game keeper, game warden
Henne	hen
Heringsmöwe	lesser blackbacked gull
Hermelin, Großes Wiesel	stoat, ermine
Herz	heart
Hetze, Hatz	hunting with hounds, pack of hunting dogs, forced hunt
Heuler	young seal
Hexenring	a circle of footprints of a doe and buck on heat
Hexensteig	witches path
Himmelsziege, Bekassine, Schnepfe	snipe
Hinterlauf	hind leg
Hinterpfote	hind paw
Hinterschaft	buttstock
Hirsch, Rothirsch	red deer, hart, stag
Hirschbrunft	rut of deer
Hirschdecke	skin of stag
Hirschfänger	cutlass, dagger
Hirschgeweih	stag´s antlers, horns
Hirschhund	staghound
Hirschkalb	fawn, veel, male calf of red deer
Hirschruf	deer call, instrument for roaring in the rutting time
Hitze, Ranz, Ranzzeit, Rollzeit	heat, period of heat
hochbeschlagen, innehaben, dickgehen, trächtig	pregnant
hochblatt	high on the chest
Hochhaus	high house
hochmachen	to rouse, to put up

Hochschuss	high shot
Hochsitz, Kanzel	high seat
hochwerden	to rise
Hochwild, Großwild	large (big) game
Hochwildjagd, Großwildjagd	big game stalking, big game hunting
Hochwildkaliber	big game calibre
Hochzeitskleid, Prachtkleid	nuptial plumage, nuptial coloration
Höckerschwan	mute swan
Hohe Jagd	big game shooting
Hohle	cavity, sole
Höhle	cave, cavern, cavity
Höhlenbrüter	cave breeder
Hohlspitzgeschoss	hollow point bullet
Hohltaube	stock dove
Holle (Federbüschel)	crest, tuft
Holzbock, Zecke	tick
hoppeln	to hop
Horn	horn
Hörner, Federohren	plumicorn, feathered ear, ear tuft
Horst	nest, aerie, eyrie, hurst
horsten	to build an eyrie
Hose (Federn an den Ständern mancher Greifvögel und Raufußhühner)	flag, thigh
Hubertustag	St. Hubert´s day
hudern	to gather under the wing
Hühnergalgen	bird sling, gallows for partridges
Hühnerhabicht, Habicht	goshawk
Hühnerhund, Vorstehhund	pointer, pointing dog
Hühnerjagd	partridge shooting
Hülse	cartridge
Hundebandwurm	dog tape-worm
Hundeführer	doghandler
Hundehalter	dog owner

Hundehütte	dog house
Hundeleine	dog lead | dog leash
Hundemeute	pack of hounds
Hundepfeife	dog whistle
Hunderasse	dog breed
Hundestaupe	canine distemper
Hundezüchter	dog breeder
Hundezwinger	kennel
Hündin	bitch, female dog
Hundsrobben	earless seals
Hütte, Krähenhütte, Uhuhütte	butt, owl hut
Hüttenjagd	hut for shooting over decoy owl
Hyäne	hyaena, hyena

I

Iltis	western polecat
Impala	impala
Impfung	vaccination
innehaben, hochbeschlagen, dickgehen, trächtig	to be pregnant
Innenballistik	interior ballistics
Irish Setter	Irish setter

J

Jagd, Weidwerk	hunting, stalking, shooting, chase
Jagdanschlag	ready position
Jagdarten	hunting methods
Jagdaufseher, Heger, Jagdhüter	game keeper, game warden
Jagdausrüstung	hunting equipment
jagdbar	huntable

jagdbares Wild	game animals, animals of the chase
Jagdbeute	game, bag
Jagdeinladung	invitation for hunting
Jagderlaubnisschein	hunting licence, hunting permit
Jagdfalke	falcon, varios breeds
Jagdfasan	pheasant
Jagdfrevel, Jagdvergehen	unlawful shooting, sacrilege against hunting
Jagdführer	guide
Jagdgast	shooting-guest, hunting guest
Jagdgebiet	hunting ground (territory)
Jagdgehege	warren, enclosed pasture, paddock, enclosure, preserve, reservation
Jagdgesellschaft	hunting party
Jagdgesetz, Jagdrecht	game law
Jagdglas, Fernglas, Fernrohr, Prismenglas	binocular, field glass, telescope
Jagdgrenze	boundary
Jagdhaftpflichtversicherung	hunting liability insurance
Jagdherr (gilt auch für Pächter)	owner of shooting grounds
Jagdhorn	hunting horn, bugle
Jagdhund	hunting dog
Jagdhut	hunter´s hat, hunter´s cap
Jagdhüter, Heger, Jagdaufseher	game keeper, game warden
Jagdhütte, Jagdhaus	hunting box, hunting lodge
Jagdjahr	hunting year
Jagdleiter (Eigentümer der Meute)	master
Jagdleopard	cheetah
jagdliches Schießen	shooting practice
Jagdmesser	hunting knife
Jagdnicker	hunting knife, skinner
Jagdpacht	hunting lease
Jagdpächter	game tenant, leaseholder
Jagdpatrone	game cartridge

Jagdrecht (Recht zu Jagen)	shooting right, hunting right
Jagdrevier	hunting territory, hunting ground
Jagdschaden	hunting damage
Jagdschein	shooting-licence
Jagdschirm, Schirm	hide, screen
Jagdschutz	hunting control
Jagdsignal	signals blown by hunters during the driving
Jagdstuhl, Sitzstock	shooting seat, stick, seat stick
Jagdtasche	game bag
Jagdtrieb	hunting instinct
Jagdunfall	hunting accident
Jagdunfallversicherung	hunting accident insurance
Jagdverband	hunting association
Jagdvergehen, Jagdfrevel	unlawful shooting, sacrilege against hunting
Jagdverhalten	hunting behaviour
Jagdwaffe	game gun, gun of a hunter, weapon
Jagdwagen	hunting carriage
Jagdwild	game
Jagdwissenschaft	wildlife research
Jagdzeit, Schusszeit	shooting season, hunting season
jagen	to hunt, to pursue, to chase, to shoot
Jäger, Erleger, Schütze, Weidmann	shooter, hunter, gunner, marks(wo)man (huntress)
Jägerlatein	hunters yarn
Jägerprüfung	hunters exemination
Jägerrecht, Aufbruch	bowels
Jägersprache, Weidmanns-sprache	hunters terms, hunters slang
Jägersteig, Pirschweg, Pirschsteig, Pirschpfad	path, hunting trail

Jaguar	jaguar
Jahresjagdschein	annual hunting licence
Jährling	yearling buck (second year)
Jule	crutch-pole, perch
Junge	juvenile, young animal
Jungfuchs, Fuchswelpe	fox cub
Junghase	leveret
Jungwild	young game

K

Kahlhirsch, Mönch, Plattkopf	antlerless stag
Kahlwild	female and young (hornless) deer (red, fallow) and elk, hind
Kaiseradler	imperial eagle
Kalb	calf
Kaliber	calibre, bore
kalte Fährte	cold trail
kalte Waffen, blanke Waffen	cold steel, cold weapons (hunting knife)
Kamel	camel
Kamm	comb
kämmen	to comb
Kammer	bolt
Kammerschuss	heart shot
Kammerstengel	bolt lever
kämpfen	to fight
Kampfläufer	ruff
Kampfruf, Sprengruf	fighting sound, charging sound
Kampfsprosse	fight-tine
Kanadagans	Canada goose
Känguru	kangaroo
Kaninchen	rabbit
Kaninchenbau	rabbit burrow (hole)

Kaninchenhaube, Frettiernetz, Netz	purse net
Kaninchenpest, Myxomatose	myxomatosis
Kanzel, Hochsitz	high seat
kapital	capital
Karakal	caracal
Karibu, Rentier	caribou (amerik.), reindeer
Kastenfalle	box trap
Kater	tom (male) cat
Katze	cat, she-cat
Katzenluchs	lynx
Kauz	owl
keckern	to snarl, to chatter
Kegel (Männchen machen)	skittle
Kegelrobbe	grey seal
Kehle	throat, gula, gullet
Kehlkopf, Drosselknopf, Stimmknoten	larynx
Keiler	wild boar, tusker, boar
Kelle, Biberkelle	beaver´s tail, beaver slice
Kern (abgebalgter Raubwilkörper)	kernel, carcass
Kessel	kettle
Kesseljagd	circular beat (hunt)
Kesseltreiben	circular drive, circular beat
Kette	chain, covey of birds, team of ducks
Keule	clava, club, haunch, hind leg, thigh
Kiebitz	lapwing (plover)
Kiefer	jaw, jawbone
Kieferast	maxillary branch, ramus mandibuli
Kimme	rear sight notch, sighting notch
Kipplaufbüchse	breakaction rifle
Kipplaufwaffe	scattergun, break-down gun

Kirchgang	going to church
kirren, ankirren, anködern, ludern	to put food for the game, to lure with a bait (carcass), to bait, to decoy, to entice, to lure, to attract
Kirrung	attractant, bait
Kitz	kid, fawn, roe kid
Kitzgeiß	kid (female) of ibex, chamois, deer
Kitzruf	calling sound of a kid, cry of fawncalling sound of a kid
klagen	anxious call, to whine
Klapper	rattle
klappern	to rattle
Klappmesser	flick-knife
Klaue	claw
Kleiner Münsterländer	Small Münsterländer
Kleinkaliber	small bore, small caliber
Klinge	blade
Knäkente	garganey
knappen, schnalzen, zählen	clicking note
knautschen	to pinch
Knieper	fallow-buck of the second head
Knochen	bone
Knochensplitter	splinter of bone, bone fragment
Knopfbock	roe buck with knobs instead of horns, button spiker
knören, trensen	short bellow
Köder	bait, lure
ködern, ankirren, anludern	to decoy, to entice, to lure, to bait, to put food for the game, to lure with a bait (carcass)
Kohlfuchs	black fox
Kojote	coyote, praire wolf, barking wolf
Kokzidiose	coccidiosis

Kolbenente	red-crested pochard
Kolbenhals	wrist
Kolbenhirsch	stag with velvet-covered antlers
Kolkrabe	raven
Koloniebrüter	colonial nester
kombiniertes Gewehr	combination-barrelled gun
Kondor, Kalifornischer	california condor
Königsadler, Steinadler	golden eagle
Kopf, Geweih, Grind, Haupt	antlers head
Kopfhirsch	leading stag
Koppel	couple
Korkenzieher	corkscrew
Kormoran	cormorant
Korn	foresight, bead
Kornweihe	hen harriers
Kotfresser	coprophage
Kragen, Brunftmähne, Mähne	rutting-mane, mane
Kragenbär	tibetan black bear
Krähe	crow
Krähenfuß	crow-foot
Krähenhütte, Hütte, Uhuhütte	butt, owl hut
Krähennest	crow´s nest
Krähenscharbe	shag
Krammetsvogel (Wacholderdrossel)	fieldfare
Kranich	crane
krank	sick, ill, diseased
Krankheit	illness, sickness, disease
krankschießen, anschießen, anschweißen	to wound, to injure
kreisen	to track, to circle
krellen, federn	to stun, to hit the spine
Krellschuss	shot at the spine
Kreuzbock	stag with cruciform antlers
Kreuzfuchs	„cross" variety of the red fox
Krickel, Gamskrucken, Krucken	chamois horn

Krickente	teal
Krone	crown
Kropf	craw
Kropf (für Falken)	crop, upper breast
kröpfen	to feed
Krucken, Gamskrucken, Krickel	chamois horn
Kruppe	crupper, croup
Kuder	male cat
Kudu, Großer	(greater) kudu
Kudu, Kleiner	lesser kudu
Kugel, Geschoss	bullet
Kugelfang	bank of earth, backstop, butt
Kugelschlag	telling
Kuhantilope	hartebeest
Kuhreiher	cattle egret
kullern	to gobble
Kulturflüchter	animal avoiding areas of human activity
Kulturfolger	animal settling in cultivated areas
kupieren, coupieren	to cut a tail, to dock, to crop
Kurzschnabelgans	pinkfooted goose
Kurzwildbret, Brunftkugel	coddes testicle

L

Labrador-Retriever	Labrador-retriever
Lachmöwe	black-headed gull
laden	to charge, to load
Lagerfeuer	campfire
Lagerschnepfe	snipe which is staying the whole year (also in winter) in the home location
Lamm, Schaflamm	lamb of moufflon
Landespflege	landscape management
Landschaftsschutz	landscape protection
Landschaftsschutzgebiet	protected landscape
Langfessel, Fessel	leash
Langhaar	long-haired hunting dog
Langhaarteckel	long-haired dachshund
Lannerfalke	lanner falcon
Lappen	toils
Latschen, Ruder	palmate foot, webbet foot
Lauf 1) Waffe, 2) Bein	1.) barrel, 2) leg (of animal)
läufig	on heat, ruttish
lauflahm	lame of a leg
Laufmündung	muzzle
Laufschiene	rib
Lauscher, Luser	ear
Laut	sound, to give tongue
Lautgeben des Jagdhundes, anschlagen,bellen, verbellen	to give tongue (hunting dog), to bark.
Lavarack, English Setter	English setter
Lebensraum	habitat
Leber	liver
Leberegel	liver fluke
Lecker, Graser, Weidlöffel, Zunge	tongue
Leckstein, Lecke, Salzlecke	salt lick
leer	empty
Lefzen	flews, lips

leg dich!	down
legen	to lay
Leier, Schar, Schere, Spiel	lyre shaped tail, tail, playing tail
Leitbache	leading sow
Leiter, Ansitzleiter	ladder
Leithund	lime-hound
Leittier	leader of a herd, leading animal
Lemming	lemming
Lende	loins, hip
Leopard	leopard
Lerchenfalke, Baumfalke, Schwalbenfalke	hobby
Leuchtpunkt	illuminated point
Licht (Auge)	eye (of large game)
liegen	to lie
Limikolen	shore birds
Lippenbär	Indian bear
Lizenz	licence
Lockente	decoy duck
Locker, (auch Wildlocker)	animal calling
Lockjagd	calling
Lockmittel	lure
Lockruf	call note
Lockvogel	decoy bird
Löffel, Hasenohr	hare´s ear, spoon (fiction)
Löffelente	shoveler
Löffler	fallow buck of the third head
lösen	to dung, to drop
Losung	dung, droppings, fewnet,
Löwe	lion
Löwin	lioness
Luchs	lynx
Luchsin	female lynx
Luder, Fallwild	carrion
ludern	to lure with a bait

Luderplatz	place of carrion, bait place
lüften	to ventilate, to air
Luftröhre, Drossel	trachea, windpipe
Lunge	lung
Lungenwurm	lung worm
Lunte, Fahne, Rute, Standarte	breech, brush, drag, tail
Luser, Lauscher	ear

M

Magazin	magazine
Magen, Bauch	belly, stomach
Magensteine	gastrolith, gastric calculus
Mähne 1) Löwenmähne, 2) Brunftmähne, Kragen	1) lions mane, 2) mane, rutting mane
mahnen	to call
Mähnenwolf	maned wolf
Mais	maize
Malbaum	a tree on which hoofed game rubs
Malerfeder, Schnepfengrandel	pin feather
Mammut	mammoth
Männchen (Federwild)	cock, male (for feathered game)
Männchen machen, Kegel	skittle
Mantel	pallium, mantle
Mantelmöwe	great black-backet gull
Mantelpavian	hamadryas baboon
Marco-Polo-Schaf	Marco Polo argali
Marder	marten
Marderhund	raccoon dog
markieren	to mark
Maske, Zügel	mask
Mast	mast
Mastbaum	seed tree
Maulkorb	muzzle

Mäusebussard, Bussard	buzzard
mäuseln	to imitate mouse calls
Mauser, Federwechsel, Raue	moult, moulting
Mausererpel, Rauerpel	moult drake
mausern, schildern	to moult
Mauspfeifchen	mouse call, mouse pipe
Mauswiesel	weasel
meckern	the call of moufflon, to bleat
Meckern der Bekassine	bleating with the wings (for becassine, snipe)
melden, orgeln, röhren, schreien	to bell, to roar
Merlin	merlin
Meute, Saumeute	pack of hounds
Meuteführer, Piqueur,	huntsman
Milchgebiss	milk dentition, primary dentition
Milz	spleen
Milzbrand	anthrax
Mink (auch Nerz)	american mink
mitnehmen	to follow a fleeting animal
mitschwingen	to follow through
Mittelsäger	red-breasted merganser
Mittelsprosse	tray
Mitteltrieb	adjustment wheel
Moderhinke	foot rot
Molar, Backenzahn	molar, molar tooth, backtooth
Mönch, Kahlhirsch, Plattkopf	antlerless stag
Mönchsrobbe	monk seal
monogam	monogammous
Monogamie	monogamy
Moorente	ferruginous duck
Moorhuhn	grouse
Moorschneehuhn	willow grouse
Moorschneehuhn, Schottisches	red grouse
Mörder	murderer
Morgenansitz	still hunt at dawn
Moschusbock	musk beetle

Moschushirsch	musk deer
mucken (Schuss verreißen)	to flinch
Mückenschleier	mosquito net
Muffelschnecke, Schnecke, Schlauch	horn of moufflon
Muffelwild, Mufflong	moufflon
Mufflon, Muffelwild	moufflon
Mündungsschoner	muzzle cover
Munition	ammunition
Muntjak	muntjac
Murmeltier	marmot
murren	to snarl
Muttertier, Mutterwild	female deer
Myxomatose, Kaninchenpest	myxomatosis

N

Nachbalz	postmating
Nachbrunft	after-rut
nachhängen	to follow with distance
nachladen, wiederladen	to reload
nachprellen	to persue
Nachsuche	trailing, searching with a hunting dog for wounded game, following a wounded game
nachtaktiv	nocturnal
Nachtglas	night-glass
Nackenwind	back-wind
Nagel	claw, nail
nageln	to nail
Nager	rodent
Näpfchen	bowl
Nase, Windfang	nose
Nasenbär	coati

Nasenbein	nasal bone
Nashorn	rhinoceros
nässen, feuchten	to piss, to urinate
Naturgesetz	law of nature
Naturschutz	nature conservation
Naturschützer	conservationist
Naturschutzgebiet	nature preserve, conservation area
Naturtrieb	instinct, natural urge
Nebel	fog, mist
Nebelkrähe	hooded crow
Nelke, Brunftfeige, Viole, Duftdrüse	scent gland
Nerz, Amerikanischer Nerz, Mink	american mink
Nerz, Europäischer	European mink
Nest	nest
Nestbau	nesting
Nestflüchter	precocial bird
Nesthocker	nestling, nidicolous bird
Nestling	nestling
Neue, Neuschnee, Spurschnee	recent snow
Nicker	hunting knife, skinner
Niederhaus	low house
Niederjagd	low game shooting
niedertun	to lie down, to bed down
Niederwild	small (low) game
Niederwildjagd	shooting of low game
Niederwildjäger	small game hunter
niederziehen	to pull down
Niere	kidney
Nilpferd	hippopotamus
nisten	to nest
Nisthöhle	nesting hole
Nistkasten	nesting box
Nistplatz	nesting place
Nonnengans, Weißwangengans	barnacle goose

Notbau, Fluchtbau	burrow for taking flight, emergency burrow
Nutria, Biberratte, Sumpfbiber	coypu, nutria
Nutzwild	productive (eatable) game

O

Oberkiefer	maxilla, upper jaw
Oberrücken, Afterklauen, Geäfter	hind claws, dew claw, ergot
Objektiv	objective lens
Offene Visierung	open sight
Ohrenzwang	ear-ache
Ohrschützer	ear flap
ökologisch	ecological
Okular	ocular lens
Orang-Utan	Orang-utan
orgeln, melden, röhren, schreien	to bell, to roar
Oryxantilope	Oryx gazella
Otter (männlicher Otter)	otter (otter dog)
Otterin (Fähe)	bitch otter
Ozelot	ocelot

P

Paar	pair
paaren	to copulate, to mate, to pair
Paarhufer, Schalenwild	cloven-hoofed animal, hoofed game
Paarungszeit	mating season
packen	to seize
Packer, Finder, Saufinder	boarhound, tracker, trailer

Pansen, Weidsack	paunch, rumen, stomach
Panseninhalt	rumen content
Panther	panther
Panzer	shield
Panzernashorn	great Indian rhinoceros
Parasit	parasite
Parforcejagd	hunting par force
Parson-Jack-Russell-Terrier	Parson-Jack-Russell-Terrier
Pass	path
Patrone	cartridge
Patronenetui	cartridge bag
Patronengürtel	cartridge belt
Patronenhülse	cartridge case
Patronenlager	chamber, magazine
Patronenmagazin	magazine
Pavian	baboon
Pelzjäger	trapper
Pendelstange	pendulum beam
Perlen	pearls
Perückengeweih	wig antlers
pfeifen	to whistle
Pfeifente	wigeon
Pfeil	arrow
Pferdeantilope	roan
Pfote	paw
Pfropfen, Zwischenmittel	wad
Pinsel	tuft of hair, brush
Piqueur, Meuteführer	huntsman
Pirsch, Birsch, Pürsch	stalking
pirschen, pürschen	to stalk
Pirschjäger	stalker
Pirschpfad, Pirschsteig, Pirschweg, Jägersteig	hunting trail, path
Pirschzeichen	mark, sign, spoor, track
Pistolengriff	pistol grip
Plattkopf, Kahlhirsch, Mönch	antlerless stag
Platzbock	place roe buck

plätzen	scraping with the fore legs, to paw
Platzhirsch	place stag
Pointer	pointer
Polarfuchs, Eisfuchs	arctic fox
polygam	polygamous
Population	population
Posten (Rehposten)	buckshot
Prachtkleid, Hochzeitskleid	nuptial plumage, nuptial coloration
Pranke, Branke, Brante, Prante	claw, clutch, pad, paw
Präparator	preparator
Präriewolf	coyote
Prismenglas, Fernglas, Fernrohr, Jagdglas	binocular, field glass, telescope
Probeschießen	check the zero
Pudelpointer	poodle pointer
Pufferpatrone	dummy cartridge
Pulverladung	powder
Puma	puma, mountain lion
Pürsch, Birsch, Pirsch	stalking
pürschen, pirschen	to stalk
Pürschzeichen	mark, sign, spoor, track
Pürzel, Bürzel, Federlein (Schwarzwild)	tail (wild boar)
Puter	turkey cock
Putzstock	cleaning rod

Q

Quäke	hare call
Quarthase (1 Monat)	leveret (1 month)
Quaste	tuft of hairs, tassel
Querschläger	ricochet hit

R

Rabenkrähe	carrion crow
Rachen	pharynx
Rachenbremse	throat bot fly
Rackelwild	cross between capercaillie and black grouse
Rallen	rails
rammeln	to buck
Rammelzeit	bucking
Rammler	male hare
ranzen	to rut
Ranzzeit, Hitze, Ranz, Rollzeit	heat, period of heat
Rappenantilope	sable antelope
rasant	flat
Ratte	rat
Räuber	predator
Raubvogel	bird of prey
Raubwild	predaceous animal
Raubwürger	great grey shrike
Raubzeug	vermin, pest
rauchlos	smokeless
Räude	mange, scab
Raue (Mauser – Wasserfederwild)	moult (for wildfowl), moulting
Rauerpel, Mausererpel	moult drake
Raufußbussard	rough-legged buzzard
Raufußhuhn, Schottisches Moorschneehuhn	grouse

Raufußkauz	Tengmalm´s owl
Raufutter	roughage
rauhaarig	hirsute
Rauhaarteckel	rough-haired dachshund
rauschen	to rut, to brim
Rauschzeit	heat period of wild boar, rutting season, brimming time
Rebhahn	cock partridge
Rebhenne	hen
Rebhuhn	partridge
Regenmantel	coat
Reh	roe
Rehbock (Bock)	roebuck
Rehkitz, Rickenkitz, Geißkitz	doe kid, roe kid
Rehlocker, Blatter	caller, roe deer call
Rehwild	roe deer
reif	mature, ripe
reihen	to mate, to pair
Reiher	heron
Reiherente	tufted duck
Reihzeit	mating season of ducks
reißen	to kill
Reißzahn	carnassial (canine) tooth
reizen	to stimulate, to exite, to irritate
Reizjagd, Fuchsreizen	decoying a fox, fox calling
Remise, Hegebusch	grove, copse, small woods in the landscape for game shelter
rennen	to run
Rentier, Karibu	reindeer, caribou (amerik.)
Repetierbüchse	repeater, repeating rifle
repetieren	to repeat
Revier	territory
revieren	to search with a hunting dog
Reviergrenze	territorial boundary
Revierkampf	territorial fight
Reviermarkierung	marking of territory

Revolver	revolver
Ricke, Geiß	doe, female roe, goat
Rickenkitz, Geißkitz, Rehkitz	doe kid, roe kid
riechen	to smell
Riefe (rinnenförmige Vertiefung auf der Oberfläche von Geweihen)	gutter of antler
riegeln	to beat for chamois
Riemen	leash, strap
Ring	ring, neckring (cock pheasant)
Ringelgans	brent goose
Ringelrobbe	ringed seal
Ringeltaube	wood pigeon
rinnen	to swim
Rippe	rib
Riss	carcass of game killed by predaceous animal
Röhre	entrance to the burrows of fox or badger, gallery in burrows
röhren, melden, orgeln, schreien	to bell, to roar
Rohrweihe	marsh harrier
rollieren, sich überschlagen	to somersault
Rollzeit, Hitze, Ranz, Ranzzeit	period of heat, heat
Rose: 1) Perlenwulst am unteren Stangenende der Geweihträger, 2) Balzrosen (für Raufußhühner und sonstiges Federwild), auch Flamme	1) ring at the base of antler, coronet, burr. 2) red spot above eyes of winged game, eye wattle, red wattle
Rosenstock	antler base, pedicle
Rostgans	ruddy shelduck
Rote Liste	Red Data Book
Roter Milan, Gabelweihe (zoologisch unkorrekt)	red kite
Rotfährte	red track
Rotfuchs	red fox

Rothabicht	young goshawk
Rothirsch, Edelwild, Rotwild	red deer
Rothirschgeweih	antlers
Rothuhn	red-legged partridge
Rotte	pack (of wild boars)
Rottier, Alttier, Tier	adult animal, female adult animal, female red deer, hind
Rotwild, Edelwild, Rothirsch	red deer
Rotwildkalb	red deer calf
Rücken, Ziemer	back, saddle, dorsum
Rückenwind	back wind
Rückfährte	back track
Rucksack	rucksack, backpack
Rückstecher	single set trigger
Rückstoß	kickback
Rückwechsel	back run
Rüde	dog, male of canidae
Rudel	herd, pack, troop
Ruder, Latschen	palmate foot, webbed foot
Ruf	call, cry
Rupfung	pluck
Rüssel	trunk, manubrium
Rute (Hund, Haarraubwild) Brunftrute, Feuchtglied, Fruchtglied (Hund, Haarraubwild)	penis, pizzle (predaceous animals)
Rute, Fahne, Lunte, Standarte	breech, brush, drag, tail
rutschen	to slide, to move
rütteln	to jog, to jolt, to wag

S

Saatgans	bean goose
Saatkrähe	rook
Säbelantilope	scimitar – horned oryx
Safari	safari
Salzlecke, Sulze, Leckstein	salt lick
Sammelplatz	meet
Sasse	resting place of hare, lair of hare, form
Sattel, Schabracke, Schneefleck	saddlecloth, white back of certain mammals (ram)
Sattelrobbe	harp seal
Satz	litter (hare and rabbit)
Satzzeit, Setzzeit	time for giving birth
Sau	sow, wild pig
Saubart, Federn (für Schwarzwild), Borsten	bristles (back of wild boar)
Saufang	boar-trap
Saufeder	boar spear, hunting lance
Saufinder, Finder, Packer	boarhound, trailer, tracker
säugen	to suckle, breast-feed, to nurse
Säuger	mammal
Saugloch, Schmalzröhre, Stinkloch	pocket of gland, venturi tube
Saumeute, Meute	pack of hounds
Schabracke, Sattel, Schnee-fleck	saddlecloth, white back of certain mammals (ram)
Schädel	skull, cranium
Schadhirsch	stag with branchless antlers
Schaf	ewe, sheep
Schaflamm, Lamm	lamb of moufflon
Schaft	shaft, stock
Schaftkappe	butt plate
Schakal	jackal
Schale	hoof, cloven hoof

schälen	to peel
Schalenabdruck	track
Schalenrand, Sensel	wall
Schalenwild, Paarhufer	hoofed game, cloven-hoofed animal
Schalldämpfer	silencer
Schälschaden	stripping damage
Schar, Leier, Schere, Spiel	lyre shaped tail, tail, playing tail
scharf sein (mutiger Hund)	ready to attack (courageous dog), keen
schärfen, aufschärfen	to gut or to cut up the skin of furred game or the plumage of winged game, to blade up
Scharwild, Geraffel (Gamsgeißen, deren Kitze und geringe Böcke)	herd of chamois, their kids and bucks
Schaufelgeweih	palmate antlers, palm
Schaufelkrone	palmed head
Schaufler	deer with palmed antlers
Schauflerhaken, Dorn	the tine on the ground of the shovel antler, back antler
Scheibe, Wurf	snout, rowel
Scheide	sheath
Schelladler	spotted eagle
Schellente	goldeneye
Schere, Leier, Schar, Spiel	lyre shaped tail, tail, playing tail
scherzen (Spielen der Hirsche mit ihrem Geweih, Kraft erproben)	to sport (stags play with their antlers)
schieben	to grow
Schießbahn	firing lane
schießen	to shoot
Schießer, Aasjäger	pothunter
Schießlehre	study of shooting
Schießstand	shooting range
Schild (1), 2) Blatt, Schulterblatt	1) shield (capercaillie, hazel cock, cock pheasant, partridge). 2) shoulder blade (wild boar)

schildern, mausern	to moult
Schirm, Jagdschirm	hide, screen
Schlafbaum	roosting tree
Schlafplatz	sleeping (roost) place
Schlagbolzen	striker, firing-pin
schlagen	1) to beat antlers (stag) against trees. 2) to attack dogs (wild boars) with tusks. 3) to catch (predaceous animal), to capture
Schlagstücksicherung	hammer safety
Schlangenadler	short-toed eagle
Schlauch, Schnecke, Muffelschnecke	horn of moufflon
schlecht	bad
Schleiereule	barn owl
schleifen, wetzen (Auerhahn)	to scream, to whet
Schleppe, auch Geschleife (künstliche Wundfährte für den Hund)	scented trail, trail drag, scent bag
schliefen	to crawl
Schlinge	snare
Schloss: 1) Beckenknochenverbindung, 2) Gewehrschloss)	1) pelvic joint (cloven hoofed game), 2) lock (gun)
Schlosstritt	trace of stag in the middle of the lair, where the stag was sitting
Schlund	oesophagus, gullet, throat
schlüpfen, ausfallen	to hatch, to moult, to come out
schmälen, schrecken	to bleat
Schmalreh	roe doe (in the second year)
Schmalschaf	femal mouflon in the second year
Schmalspießer	stag with first antlers, pricket, spiker
Schmaltier (2. Lebensjahr)	jinnock, hearse
Schmalz, Dachsfett, Weißes	fat of badger
Schmalzröhre, Saugloch, Stinkloch	pocket of gland, venturi tube

schmeißen (Ausscheiden der Exkremente aller Greifvögel und Reiher)	to throw (excrements of predatory birds)
Schnabel (Auerwild: Brocker)	bill, beak, lip, rostrum, spout
Schnabel (für Falken)	beak
Schnalle	vagina (dogs, predaceous animals)
schnallen, abhalsen	to let loose, to unleash, to uncouple
schnalzen, knappen, zählen	clicking note
Schnatterente	gadwall
Schnecke, Muffelschnecke, Schlauch	horn of mouflon
Schneefleck, Sattel, Schabracke	saddlecloth, white back of certain mammals (ram)
Schneegans	snow goose
Schneehase	mountain hare, blue hare
Schneehuhn	ptarmigan
Schneeleopard	snow leopard
Schneeziege	Rocky Mountain goat
Schneide	cutting edge
Schneider	weak stag, young cock for capercaillie and black cock
Schneidezahn	cutting tooth, incisor
Schneise, Waldschneise	aisle, vista
Schnepfe, Bekassine, Himmelsziege	snipe
Schnepfenbart	hair dot, beard of snipe
Schnepfengrandel, Malerfeder	pin feather
Schnepfenstrich	passage of snipes
Schnitthaar	hair of a wounded game after shooting
schnüren	footmarks on a straight line (fox, lynx, wild cat, wolf)
schonen: 1) hegen, 2) Nichtaufsetzen eines Laufes nach Laufverletzung	1) to preserve, 2) a game with wounded legs that doesn´t put all foots on the ground

Schonzeit	close (closed) season
Schonzeitbüchse	closed-season rifle
Schoof (Schof) für Enten	flock (family of ducks)
Schopf	tuft (heron, tufted duck, goosander)
schöpfen (trinken)	to drink
Schrank, schränken	cross trace
schrecken, schmälen	to bleat
Schreiadler	lesser spotted eagle
schreien, melden, orgeln, röhren	to bell, to roar
Schritt	step
Schrittlänge	stride
Schrot, Hagel	small shot
Schrotgarbe	shot column
Schrotladung	shot charge
Schrotpatrone	shotgun cartridge
Schulterblatt, 1) Blatt, 2) Schild	1) shoulder blade (wild boar), 2) shield (capercaillie, hazel cock, cock pheasant, partridge)
Schürze	hair dot
Schuss auf flüchtendes Wild	running shot
Schüssel, Teller	ear of wild boar
Schüsseltreiben	dinner after hunting
Schussentfernung	shot distance
schussfest	shot-proof
schussscheu	gun-shy
Schusswaffe	firearm
Schusszeichen	shot mark
Schusszeit, Jagdzeit	shooting season, hunting season
Schütte	feed trough, feed hopper
Schütze, Erleger, Jäger, Weidmann	hunter, marksman, marks-woman, shooter, gunner
Schützenbruch	branch for a successful hunter
schwach, gering	of poor condition, weak, little, small

Schwalbenfalke, Baumfalke, Lerchenfalke	hobby
Schwan	swan
Schwanenhals	jump trap
Schwanz (für Falken)	tail
Schwarte 1) Dachs, 2) Schwarzwild	1) coat, pyles, skin, 2) hide, pyles
Schwarzbär	black bear
Schwarzstorch	black stork
Schwarzwild	wild boar
Schwarzwildlosung	lesses
Schweinepest	swine fever
Schweiß	blood (game)
Schweißarbeit	blood-trailing
schweißen, ausschweißen	to bleed
Schweißfährte, Wundfährte	blood trail, red track
Schweißhund	bloodhound
Schweißriemen	long leash
Schwinge, Flügel	pinion, wing
Sechser, Sechsender	six points stag
Seeadler	white-tailed eagle
Seebär	fur seal
Seehund	harbour seal
Seelenachse	line of bore
Seeotter	sea otter
Seher	eyes of small game
Sehschärfe	visual power
Selbstlader	autoloading shotgun
Senkung (Schaftsenkung)	drop
Sensel, Schalenrand	wall
Setter	setter
setz dich!	down!
setzen	to calve (deer), to kid, to lamb, to cub (carnivore), to litter (to farrow, wild boar), to kindle (hare, rabbit)
Setzhase, Häsin	female hare

Setzzeit, Satzzeit	time for giving birth, fawning season
Sichel (gekrümmte Stoßfedern des Birkhahns)	tail of black-cock, curved tail
sichern: 1) verhoffen, 2) Waffe	1) to wind, to stop, 2) set at safe
Sicherung, Sicherheit	safety
sichtlaut	hunt by sight
Sikahirsch	sika deer
Silberfuchs	silver fox
Silbermöwe	herring gull
Singschwan	whooper swan
Sitzstock, Jagdstuhl	shooting seat, stick, seat stick
Skeetschießen	skeet shooting
Sonnenbalz	courtship after sunrise (black grouse)
Sonntagsjäger	holiday sportsman
Spaniel	spaniel
Spannhebel	cocking lever
Spektiv	telescope
Sperber	sparrow hawk
Spiegel (Auer- und Birkwild)	wing patch (capercaillie, black grouse)
Spiegel (Ente)	mirror (duck)
Spiegel (Schalenwild)	target, buff patch, rump patch (hoofed game)
Spiel, Leier, Schar, Schere	lyre shaped tail, tail, playing tail
Spiel, Stoß (Fasanenhahn)	tail (cock pheasant)
Spielhahn, Birkhahn	blackcock
Spießbock, Spießer (gilt auch für Rotspießer)	spike stag, spiker
Spießente	pintail
Spießer	pricket, spiker, spike stag
Spinne, Gesäuge	udder, mammary glands of animals
spitz	head on
Sporen, Sporn	spur (pheasant)

sprengen	to blast
Sprengruf, Kampfruf	charging sound, fighting sound
Springbock	sprinbok,springbuck
springen 1) Verlassen des Baues durch Baubewohner, 2) Springen aufs Blatt (Rehbock)	1) to jump (animals, who are living in burrows), 2) to jump after calling, or bleating (roe buck in the rutting season)
Springer Spaniel	springer spaniel
Sprinz	sparrow hawk (male)
Sprosse, Ende, Geweihzacken	branch, prong of antler, point, tine
Sprung (Rehwildverband)	lead of roe deer
Sprünge	hind legs (hare)
Spur, Fährte, Wildspur	trail, spoor, track
spüren, abfährten, abspüren	to track, to follow the trail of animals
spurlaut	speaking while trailing
Spurschnee, Neue, Neuschnee	recent snow
Stachelschwein	porcupine
Stammbaum	pedrigree
Stand	shooting position
Standarte, Fahne, Lunte, Rute	breech, brush, drag, tail
Ständer	leg, shank, tarsus
Standlaut	standing at bay, continuous barking
Standtreiben	driving
Standwild	resident game
Stange, Geweihstange	beam
Stangenlänge	length of the antlers
Stangensicherung	sear safety
stark	strong
Staupe	dog distemper
stechen, einstechen	to prick, to shoot with a hair-trigger
stechen, wurmen (Schnepfe)	to search for food (snipe), to pierce, to sting, to worm
Stecher	bill of snipe

Stecher, deutscher Stecher	double set trigger, hairtrigger
stecken (Sauen)	to stick (wild boars stay in underwood)
stehen: 1) Einstand, 2) weibliches Schalenwild steht dem männlichen Stück beim Beschlag, 3) vorstehen	1) to stay in a certain district (game), 2) to stay for a female deer, which will be covered by a stag, 3) to point 3) the hunting dog is standing in front of low game. He signals the presence ol low game
Steinadler, Königsadler	golden eagle
Steinbock	mountain goat, ibex
Steinhuhn	rock partridge
Steinkauz	little owl
Steinmarder	beech marten
stellen: 1)Stellen einer Falle 2) Wild stellen	1) to put up the trap, 2) to bay (for big game)
Steuerfedern	rectrix, rudder feathers
Stich	lower breast
Stichelhaar	rough-haired pointer
Stimmknoten, Drosselknopf, Kehlkopf	larynx
Stingel (Hals von Auer- u. Birkwild)	throat (neck of capercaillie and grouse)
Stinkloch, Saugloch, Schmalzröhre	pocket of gland, venturi tube
Stirnzapfen	bony core
Stöberhund	flushing dog
Stöberjagd	flushing
stöbern	to rummage
Stockente	mallard
Stop (Stirnabsatz)	stop
Storch	stork
Stoß, Spiel	tail, train (winged game)
stoßen	truss the quarry, to shoot down upon prey
Stoßendbinde, Endbinde	sub terminal band

Stoßzahn	tusk
strafen	to punish
Strauß	ostrich
Strecke, Beute, Jagdstrecke	bag, prey, booty, capture
strecken, erbeuten, erlegen	to kill, to bag
streichen, fliegen	to fly
Streife	walking up, hunters and beaters advance in the same line
streifen, abbalgen, abstreifen	to skin
Streifschuss	grazing shot
streuen (Munition)	to spread
Strich	flight of birds
Strichvogel	bird of passage, migratory bird
Strichzeit	time of flight
Stubendressur	indoor training
stubenrein	clean
Stück	head of game
stumpf	blunt
Stümpfe	female horn of a sheep
Sturmmöwe	common gull
Stutzen	short rifle
Suche, absuchen	tracking, dogging, to search
Suchjagd	dogging, tracking
Suhle	wallow, wallowing place
suhlen	to wallow
Sulze, Salzlecke	salt lick
Sumpfbiber, Biberratte, Nutria	coypu, nutria
Sumpfohreule	short-eared owl

T

Tafelente	pochard
tagaktiv	diurnal
Tagesjagdschein	daily hunting licence
Talg (Fett aus dem Körperinneren)	tallow (fat inside the body)
Tannenhäher	nutcracker
Tannenmarder, Baummarder	pine marten
Taube	pigeon, dove
Tauber	cock-pigeon
Tauchente	diving duck
Teckel, Dachshund, Dackel	dachshund
Teilmantelrundkopfgeschoss	softpoint round nose bullet
Teilzieher	partial migrant
Teller, Schüssel	ear of wild boar
Tellereisen	plate trap
Terzel	tercel
Tesching	pea rifle
Tier, Alttier, Rottier	adult animal, hind, female adult animal, female red deer
Tierarzt	veterinary
Tiergarten	zoologica garden, zoo
Tierkunde	zoology
Tierschutz	animal protection
Tierseuche	zoonosis, pest
Tierwelt	animal world
Tiger	tiger
Timberwolf	timber wolf
Tollwut	rabies
Tontaubenschießen	clay birding
töten	to kill
totverbellen	barking at a shot game
totverweisen	to imply (indicate) a shot game
traben	to trot (fox, badger, wolf)
trächtig, dickgehen, hochbeschlagen, innehaben	pregnant

Träger (Hals des Schalenwildes außer Schwarzwild), Hals (Niederwild)	neck (hoofed game)
Tragzeit	gestation period
Tränengrube	lachrymal fossa, tear-pit
Trappe	bustard
Trauerente	common scoter
Treffer	hit
Treffpunkt: 1) Sammelplatz. 2) Auftreffpunkt	1) meeting point. 2) point of impact
treiben, abtreiben, durchgehen, durchtreiben	to drive small game
Treiber	beater, driver
Treiberwehr	all the beaters in the beat (driving)
Treibjagd	driving, beat, battue
trensen, knören	short bellow
treten (Schalenabdruck, Trittsiegel)	footprint, track
treten: 1) austreten, 2) auswechseln (Schalenwild)	1) to tread the hen, 2) to leave or to change the home district (wood) for searching feed (for hoofed game)
Trichine	trichina
Trittsiegel, Schalenabdruck	footprint, track
trommeln	to drum
Trophäe	trophy
Trophäenschau	trophy exposition
Truthahn	turkey, turkey cock
Truthenne	turkey hen
Tüpfelhyäne	spotted hyena
Tüpfelsumpfhuhn	spotted crake
Türkentaube	collared dove
Turmfalke	kestrel
Turteltaube	turtle dove

U

überfallen	to jump
überhegen	to overpreserve
Überläufer	hogget
Überläuferbache	female hogget
Überläuferkeiler	young tusker
überschlagen (sich), rollieren	to somersault
überstellen (Auerhahn wechselt den Ast)	to jump over (capercaillie)
überwintern	to hibernate, to over-winter
Uferschnepfe	black-tailed godwit
Uhu	eagle owl
Uhuhütte, Hütte, Krähenhütte	butt, owl hut
Umschaltschieber	selector
umschlagen	to go round a thicket (copse) for checking tracks of hoofed game
umstellen, abstellen	to post the hunters and beaters around the hunting district, to surround
Ungarisch Drahthaar, Magyar Vizsla	Hungarian wire-haired pointer
Ungarisch Kurzhaar, Magyar Vizsla	Hungarian short-haired pointer
ungerade	odd, uneven
Unterhebelrepetierer	Lever-action rifle
Unterkiefer	mandible, lower jaw
unweidmännisch	unhuntsman-like
Urhahn, Auerhahn	capercaillie

V

verbeißen	to nibble sprouts of young trees and buds, to browse
verbellen, anschlagen, bellen, lautgeben	to give tongue
Verbiss, Wildfraß, Wildverbiss	damage caused by game, browsing by game, game damage
Verbissschutzmittel	deer repellent, antigame protective agent
verblasen	to sound death
verblenden	to hide
verbrechen	to mark the killed hoofed game with a branch
verecken	to furnish
verenden	to die
verendet	dead
verfärben	to change hair
verfolgen	to follow a fleeing animal
verfolgen	to hunt, to chase
vergrämen, verprellen	to disturb
Verhalten, Benehmen	behaviour
verhitzen	rotten game, to rot
verhoffen, sichern	to stop, to stand at gaze
verhören	to hear the call of stag during the rutting season or of capercailye and blackcock during the courtship
verkanten	to cant
verlorenbringen	to seek wounded game and to retrieve it
verlorensuchen	to track for wounded game
verludern	to putrefy, to rot
vernehmen	to hear
Versager	misfire
Verschluss	breech mechanism

Verschlussgehäuse	receiver
verschweigen	to be silent
versorgen	to take care of
vertraut	familiar
verweisen	to lead (hunting dog)
verwittern	to dull human scent
Vielehe	polygamy
Vielfraß	wolverine
Viole, Brunftfeige, Nelke, Duftdrüse	scent gland
Virus	virus
Visier	sight
Visierlinie	line of sight
Vogeldunst	bird shot
Vogellockpfeife	bird call
Vogelschutzgebiet	bird sanctuary
Vollchoke	full choke
Vollmantelgeschoss	full-jacket bullet
Vollmond	full moon
Vorderlauf	foreleg
Vorderpfote	forefoot, forepaw
Vorderschaft	forearm
vorhalten	to lead, speed-allowance
vorliegen (Bauhund im Fuchsbau)	to point pronely
vorstehen	to point, hunting dog is standing in front of low game
Vorstehhund, Hühnerhund	pointing dog, pointer
Vorstehtreiben	shooting driven game

W

Wachler, Gamsbart	chamois beard
Wacholderdrossel (Krammetsvogel)	fieldfare

Wachtel	quail
Wachtelhund, Bracke, Spaniel	spaniel
Waffe	the gun of a hunter, game gun
Waffen, Gewaff (Greifvögel)	claw, talon
Waffen, Gewaff, Hauer (Schwarzwild)	tusk of wild boar, weapon
Waffenbesitzkarte	firearms permit
Waffengesetz	firearms law
Waffenöl	gun oil
Waffenschein	firearms licence
wahrnehmen, eräugen	to see
Waldhase	woodland hare
Waldiltis	polecat
Waldkauz	tawny owl
Waldlemming	wood lemming
Waldohreule	long-eared owl
Waldschneise, Schneise	aisle, vista
Waldschnepfe	woodcock
Wanderfalke	peregrine falcon
Wapiti	wapiti
warm (Fährte, die Wittrung abgibt)	burning scent
Warzenschwein	warthog
Waschbär	raccoon
Wasserjagd	wildfowl shooting, waterfowl hunting
Wasserwild	waterfowl
Wechsel	pass, game crossing, runway
wechseln	to move, to change
Wechselwild	commuting game
Wedel (Pürzel für Schwarzwild)	scut, (tail for wild boar)
wehrhaft	capable of defending
Weidblatt	hanger
weidgerecht	huntsman-like
weidlaut	noisy, babbling

Weidloch	anus
Weidlöffel, Graser, Lecker, Zunge	tongue
Weidmann, Erleger, Jäger, Schütze	hunter, marksman, markswoman, shooter, gunner
Weidmannsheil	good hunting, good luck
Weidmannsprache, Jägersprache	hunters terms, hunters slang
Weidmesser, Jagdmesser	hunting knife
Weidsack, Pansen	paunch, rumen, stomach
Weidwerk, Jagd	hunting, stalking, shooting, chase
weidwund	wounded in the bowels
Weidwundschuss	bowel shot, gut shot
Weihe	harrier
Weimaraner	weimarian dog
Weißes, Dachsfett, Schmalz	fat of badger
Weißkopfseeadler	bald eagle
Weißwangengans, Nonnengans	barnacle goose
Weißwedelhirsch	whitetailed deer
Welpe	pup, whelp
werfen (gebären), wölfen	to whelp, to throw
Wespenbussard	honey buzzard
Westfälische Dachsbracke	Westphalian beagle
Wettervorhersage	weather forecast
wetzen, schleifen	to whet
Widder	ram
Widdergehörn (-geweih für Rehbock, abnorm)	ram´s horns
Widderlamm	ram lamb
Widergang	to double back
Widerrist	withers
wiederkäuen	to ruminate
Wiederkäuer	ruminant
wiederladen, nachladen	to reload
Wiesel	weasel

Wiesel, Großes, Hermelin	stoat, ermine
Wiesenweihe	Montagu´s harrier
Wild	game
Wildacker	field for feeding game
Wildbahn	hunting ground
Wildbestand, Bestand	stock of game
Wildbiologie	game biology
Wildbret	venison
Wilddichte	density of game
Wilddieb, Wilderer	poacher
Wildente	wild duck
Wilderei	poaching
wildern	to poach
Wildfauna	wild fauna
Wildfolge, Nacheile	trespass right, the right to follow wounded game
Wildforschung	game research
Wildfraß, Verbiss, Wildverbiss	damage caused by game, browsing by game, game damage
Wildfütterung	game feeding
Wildgans	wild goose
Wildgatter, Gatter	fence, holding paddock, railing
Wildhandel	trade in venison
Wildhändler	game dealer
Wildhege, Hege, Wildschutz	game keeping, wildlife management, game protection
Wildkalb	hind calf, female calf of red deer
Wildkaninchen	rabbit
Wildkatze	wild cat
Wildkunde	game science
Wildleder	suede, game-leather
Wildmarke	game tag
Wildnis	wilderness
Wildpark	game reserve, wildlife park
Wildreservat	game reserve

Wildschaden	game damage
Wildschadensverhütung	to prevent game damage
Wildschutz, Hege, Wildhege	game protection, game keeping, wildlife management
Wildschutzgebiet	wildlife reserve, game reserve
Wildschwein	wild boar
Wildseuche	game epidemic
Wildspur, Fährte, Spur	trail, track, spoor
Wildtier	wild animal
Wildtierbestand	wild fauna
Wildtierkunde	game science
Wildtransport	game transport
Wildverbiss, Verbiss, Wildfraß	damage caused by game, browsing by game, game damage
Wildwagen	game cart
Wildwechsel, Wechsel	game crossing, pass, runway
Wind	wind, flatulence
winden, Wind bekommen, Wind holen, wittern	to get wind off, to wind, to scent, to smell
Windfang, Nase	nose
Winterhaar	winter coat
Winterkleid	winter plumage, resting plumage
Winterruhe	winter restling, winter dormancy
Winterschlaf	hibernation, roosting
Wisent	European bison
wittern, winden, Wind holen	to scent, to wind, to smell, to get wind of
Witterung (Jägersprache: auch Wittrung)	scent, scented bait
Witterung (Wetter)	weather
Wolf	wolf
wölfen, werfen (gebären)	to whelp, to throw
Wolfssprosse	sur-royal, a branch under the crown

Wolle	hair, wool
Wundbett	bed for wounded game
Wundfährte, Schweißfährte	blood trail, red track
Wurf, Scheibe (Windfang)	snout
Würgebohrung, Chokebohrung	choke boring
Würger	shrike, killer
wurmen, stechen (Schnepfe)	to search for food (snipe), to pierce, to sting, to worm
Wurmmittel	vermifuge
Wüstenfuchs	fennec fox
Wüstenluchs, Karakal	caracal lynx

Y

Yak	yak

Z

zählen, knappen, schnalzen	clicking note
Zahn	tooth
Zahnformel	dental formula
Zahnwechsel	change of dentition
Zebra	zebra
Zebu	zebu
Zecke, Holzbock	tick
Zehe	toe
Zehenstifte, Balzfedern, Balzstifte	teeth of comb
Zehnender, Zehner	stag of ten points
Zeichen	mark
zeichnen	to mark, the movement of the wounded game after shooting

zerwirken	to carve, to butcher
ziehen 1) herausziehen, 2) von Schalenwild, 3) von Zugvögeln	1) to pull, 2) to move slowly (hoofed game), 3) to migrate
Zielballistik, Endballistik	terminal ballistics
Zielfernrohr	telescope sight
Zielfernrohrmontage	scope mount
Zielstock	aiming stick
Ziemer, Rücken	saddle, back, dorsum
Zobel	sable
Zug (Büchse)	groove (rifle)
Zug (Vogelzug)	migration
Zügel, Maske	mask
Zugvogel	migratory bird, bird of passage
Zündhütchen	primer
Zunge, Graser, Lecker, Weidlöffel	tongue
zurück! (Befehl für den Jagdhund)	back! (order to the hunting dog)
zurücksetzen	to set back
zusammenbrechen	to break down (for a wounded game after deadly shot)
Zwang	closed footmarks (stag)
Zwangswechel	obligate runway
Zweibein	bipod
Zwerchfell	diaphragm
Zwerggans	lesser whitefronted goose
Zwergsäger	smew
Zwergschwan	Bewick´s swan
Zwergteckel	dwarf-dachshund
Zwinger, Hundezwinger	kennel
Zwischenmittel, Pfropfen	wad
Zwölfender, Zwölfer	twelve points stag

Literaturverzeichnis

Edmund Launert — Biologisches Wörterbuch – Deutsch-Englisch • Englisch-Deutsch. Verlag Eugen Ulmer, Stuttgart. 1998.

Jürgen Schulte — in Biologisches Wörterbuch – Deutsch-Englisch • Englisch-Deutsch. Verlag Eugen Ulmer, Stuttgart, 1998.

Die Jägersprache – kurz gefasst. Verlag J. Neumann-Neudamm, Melsungen, 3. Auflage 2021.

Attila Zoltan, Joachim Köller, Istvan Miko, Antal Török, ÖdönZoltan — Bildwörterbuch der Jagd, Nimrod, Vadaszuisag. Denes Natur Mühely, 2000.

Quellen

Attila Zoltan — **Waffenkundliche Begriffe** in Bildwörterbuch der Jagd. Nimrod, Vadaszuisag. Denes Natur Mühely, 2000.

EdmundLaunert — **Englische Namensgebung für alle nicht heimischen Tierarten** in Biologisches Wörterbuch – Deutsch-Englisch • Englisch-Deutsch. Verlag Eugen Ulmer, Stuttgart. 1998.

Ulrich Sedlag — **Veterinärmedizin (Anatomie der inneren Organe, Wildkrankheiten)** in Biologisches Wörterbuch – Deutsch-Englisch • Englisch-Deutsch. Verlag Eugen Ulmer, Stuttgart. 1998.

ACHTUNG
Wende-
broschüre!

Impressum

ISBN 978-3-7888-2073-2

1. Auflage 2023
Printed in Germany

Erschienen im Ressort Jagd-Kultur im Auftrag des Verlages J. Neumann-Neudamm

c/o NJN Media AG
Unter dem Schöneberg 1
D-34212 Melsungen

info@neumann-neudamm.de
www.neumann-neudamm.de

Bildnachweis: Lizensfreie Shutterstock-Fotos.

Jürgen Schulte

HUNTING DICTIONARY

A hunting dictionary including hunter language

ENGLISH - GERMAN
GERMAN - ENGLISH

Reversible brochure

FOREWORD

The internationalization of hunting activity is prompting many hunters abroad to communicate with their foreign caregivers. The present dictionary is intended to contribute to this.

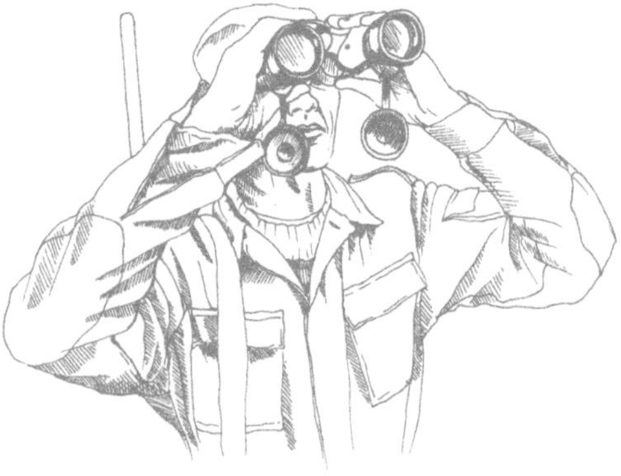

ENGLISH - GERMAN

A

abnormal, irregular	abnorm
accept (to) feeding	Annehmen der Fütterung
act of covering, copulation	Beschlag
adjustment wheel	Mitteltrieb
adult animal, female adult animal, female red deer, hind	Alttier, Hirschkuh, Rottier, Tier
adult plumage of wild fowl	Alterskleid
aerie, eyrie, hurst, nest	Horst
African elephant	Elefant (Afrikanischer)
after-rut	Nachbrunft
age class	Altersklasse
age determination	Altersbestimmung
aged, old	alt
aim (to)	anbacken, anschlagen
aim (to) by shooting on the mark 1., 2.to loose condition (game)	abkommen
aiming point	Absehen
aiming position, gun mounting, firing position	Anschlag
aiming stick	Zielstock
air (to), to ventilate	lüften
aisle, vista	Schneise, Waldschneise
alarm sound, alarm call	Angstgeschrei, Schrecklaut
alight (to) on a tree	einschwingen (Auerwild lässt sich nieder)
Alpine chough	Alpendohle
Alpine ibex	Alpensteinbock
American mink	Nerz, Amerikanischer Nerz, Mink
ammunition	Munition
ancestral register, genealogical table	Ahnentafel
animal avoiding areas of human activity	Kulturflüchter
animal calling	Locker, (auch Wildlocker)

animal protection	Tierschutz
animal settling in cultivated areas	Kulturfolger
animal world	Tierwelt
animals of the chase, game animals	jagdbares Wild
announce (to) the end of hunt by signal, to call off	abblasen
annual hunting licence	Jahresjagdschein
antelope	Antilope
anthrax	Milzbrand
antigame protective agent, deer repellent	Verbissschutzmittel
antler base, pedicle	Rosenstock
antlerless stag	Kahlhirsch, Mönch, Plattkopf
antlers, head, horns	Geweih, Grind, Haupt, Kopf, Gehörn (besser Geweih)
anus	Weidloch
anxious call, to whine	klagen
approach (to) (feathered game)	anstreichen
arctic fox	Polarfuchs, Eisfuchs
arm, file, tusk	Gewehre (untere Eckzähne des Keilers)
arrow	Pfeil
ascertain (to), to find out, to locate	ausmachen, bestätigen
assign (to) the place for hunters and beaters	anstellen
attach (to), to mount the trophies on a board	Trophäe aufsetzen
attack (to)	annehmen, angreifen
attack (to) a wild boar (hunting hounds) 1., 2. a small spread cone for a shotgun	1. decken(Saupacker greifen Schwarzwild), 2. Zusammenhalten der Schrote bei nicht streuender Flinte
attak (to) dogs (wild boars) with tusks	Verletzen der Verfolger (Hunde) von Sauen durch ihr Gewaff: schlagen

attract (to)	ködern, kirren
attractant, bait	Kirrung
aurox, wildox, urus	Auerochse
autoloading shotgun	Selbstlader

B

babbling, noisy	weidlaut
baboon	Pavian
back antler, the tine on the ground of the shovel antler	Dorn, Schauflerkaken
back of the neck, nape, nucha	Genick
back run	Rückwechsel
back tooth, molar, molar tooth	Molar, Backenzahn
back track	Rückfährte
back wind	Nackenwind, Rückenwind
back, saddle, dorsum	Rücken, Ziemer
back! (order to the hunting dog)	zurück! (Befehl für den Jagd-hund)
backer, bater	zurückgesetzter Hirsch
backpack, rucksack	Rucksack
backstop, bank of earth, butt	Kugelfang
bacterium	Bakterie
bad	schlecht
badger	Dachs
badger skin	Dachsschwarte
badger´s set	Dachsbau
bag (to), to kill	strecken, erbeuten, erlegen
bag, prey, booty, capture	Beute, Strecke, Jagdstrecke
bait (to),to decoy,to entice,to lure,to put food for the game	ködern, ankirren, anludern
bait, attractant	Kirrung
bait place, place of carrion	Luderplatz
bait, lure	Köder
bald eagle	Weißkopfseeadler

ball	Ballen
ballistics	Ballistik
bank of earth, backstop, butt	Kugelfang
bark (to), to give tongue to yelp, also by foxes during the rutting season	Lautgeben des Jagdhundes, anschlagen, bellen, verbellen, auch Lautäußerung der Füchse während der Ranz
barking (to) at a shot game	totverbellen
barking wolf, praire wolf, coyote	Kojote, Präriewolf
barn owl	Schleiereule
barnacle goose	Weißwangengans, Nonnengans
barrel	Gewehrlauf
barren	gelt, unfruchtbar
bater, backer	zurückgesetzter Hirsch
battue, driving, beat	Treibjagd
battue, hare shooting	Hasenjagd
Bavarian bloodhound	Bayrischer Gebirgsschweiß-hund
bay (to) (for big game) 1., 2. to put up the trap	1. Wild stellen, 2. Stellen einer Falle
bay (to) the dogs by stags or wild boars 1., 2. to beat off a rival by stag or wild boar. 3. to punch (hare, rabbit)	abschlagen
bay tine, bay antler	Eissprosse
bead, foresight	Korn
beagle	Beagle
beak	Schnabel (für Falken)
beam	Geweihstange, Stange
bean goose	Saatgans
bear	Bär
beard of snipe, hair dot	Schnepfenbart
beard, chamois beard	Bart
bearded vulture	Bartgeier
beat (to) antlers (stag) against trees	Abschlagen der Rinde durch den Rothirsch mit dem Geweih außerhalb der Fegezeit: schlagen

beat (to) for chamois	riegeln
beat (to) off a rival by stag or wild boar 1., 2. bay (to) the dogs by stags or wild boars. 3. to punch (hare, rabit)	abschlagen
beat of predator, hunting territory	Jagdrevier
beat, driving, battue	Treibjagd
beater shooter	Durchgehschütze
beater, driver	Treiber
beaters (all the beaters) in the beat (driving)	Treiberwehr
beating the antlers against trees by stags	schlagen
beaver	Biber
beaver´s tail, beaver slice	Kelle, Biberkelle
bed	Bett
bed down (to), to lie down	niedertun, einschieben
bed for wounded game	Wundbett
beech marten	Steinmarder
beginning of drive	antreiben
beginning of pushing the antlers	Geweih schieben
beginning of the open season	Aufgang der Jagd (Ende der Schonzeit)
behaviour	Benehmen, Verhalten
bell	Bell
bell (to), to roar	melden, orgeln, röhren, schreien
belly, stomach	Bauch, Magen
bevy, covey	Gesperre
Bewick´s swan	Zwergschwan
big game	Großwild, Hochwild
big game calibre	Hochwildkaliber
big game shooting	Hohe Jagd
big game stalking, big game hunting	Hochwildjagd, Großwildjagd

big wild boar, old solitary, singular	grobes Schwein, Hauptschwein
bill	Brocker (Schnabel Auerwild), Stecher (Waldschnepfe)
billeting, billets	Fuchslosung
binocular, field glass, telescope	Fernrohr, Fernglas, Jagdglas. Prismenglas
bipod	Zweibein
bird call	Vogellockpfeife
bird of passage, migratory bird	Strichvogel, Zugvogel
bird of prey, birds preying on game, predatory (raptorial) bird	Raubvogel, Greifvogel
bird sanctuary	Vogelschutzgebiet
bird shot	Vogeldunst
bird sling, gallows for partridges	Galgen, Hühnergalgen
bison	Bison
bitch, female dog	Hündin
bitch fox, female of canids (fox), vixen, fuchsin	Bätze, Betze, Fähe, Füchsin
bitch otter	Otterin (Fähe)
bite (to)	beißen
bite (to) young adult game by predaceous parents	abbeißen, Vertreiben ausgewachsener Jungtiere durch Raubwildeltern
black bear	Schwarzbär
blackcock	Birkhahn, Spielhahn
black coloured red fox	Brandfuchs
black fox	Kohlfuchs
black grouse	Birkwild
black-headed gull	Lachmöwe
black stork	Schwarzstorch
black-tailed godwit	Uferschnepfe
bladder, blister, bubble, vesicle	Blase
blade	Klinge
blade, shoulder blade (wild boar), shoulder 1., 2. shield (capercaillie, hazel cock, cock pheasant, partridge)	1.Blatt und Schulterblatt (Schwarzwild), 2.Schild (Auerhahn, Haselhahn, Fasan, Rebhuhn)

blast (to)	sprengen
blaze (to)	aufschärfen, schärfen
bleach (to)	bleichen
bleat (to)	schmälen, schrecken
bleat (to) (chamois)	blädern (Gamswild)
bleat (to), call of moufflon	meckern
bleat (to), to call, to whistle	blatten, fiepen
bleating with the wings (for becassine, snipe)	Meckern der Bekassine
bleed (to)	schweißen, ausschweißen
blind	blind (Gamsbart ohne Reif)
blister, bladder, bubble, vesicle	Blase
blood (game)	Schweiß
bloodhound	Schweißhund
blood of furred vermin	Farbe
blood trail, red track	Schweißfährte, Wundfährte
blood-trailing	Schweißarbeit
blow (to)	blasen
blow (to) 1. wild boar (to puff). 2. hunting horn	blasen: 1. Schwarzwild. 2. Jagdhorn
blow (to) off	abblasen
blow (to) on, to call on	anblasen
blue hare, mountain hare	Schneehase
blunt	stumpf
boarhound, trailer, tracker	Finder, Saufinder, Packer
boar spear, hunting lance	Saufeder
boar-trap	Saufang
boil (to), to scald	abkochen
bolt	Kammer
bolt lever, bolt handle	Kammerstengel
bolting, earth hunting	Baujagd, Erdjagd
bone	Knochen
bone fragment, splinter of bone	Knochensplitter
Bonelli´s eagle	Habichtsadler
bony core	Stirnzapfen
booty, capture, bag, prey	Beute, Strecke
bore, calibre	Kaliber
boundary	Jagdgrenze

bowel shot, gut shot	Weidwundschuss
bowels, entrails	Aufbruch, Jägerrecht
bowels, intestine, gut	Darm
bowl	Näpfchen
box trap	Kastenfalle
brain	Gehirn
branch for a successful hunter	Schützenbruch
branch, broken branch	Bruch
branch, prong of antler, tine, point	Ende, Geweihzacken, Sprosse
break (to) (snap) off	abbrechen
break (to) down (for a wounded game after deadly shot)	zusammenbrechen
break (to) in, to train, to train the dog for hunting	abrichten, abführen, führen
break (to) off shooting	absetzen
break-down gun, scattergun	Kipplaufwaffe
breakaction rifle	Kipplaufbüchse
breast, chest	Brust
breech, brush, drag, tail	Fahne, Lunte, Rute, Standarte
breech mechanism	Verschluss
breeding place, nesting place,	Brutplatz, Nistplatz
breeding season, brooding time	Brutzeit
brent goose	Ringelgans
brim (to), to rut	rauschen (Schwarzwild)
brimming time, heat period of wild boar, rutting season	Rauschzeit
bring (to), to retrieve, to fetch	bringen, apportieren
bristles (back of wild boar)	Borsten, Federn (Rückenhaare oder Saubart des Schwarzwildes)
broadside on	breit (Profilstellung des Wildes)
brocket, fork stag	Gabelhirsch, Gabler
broken-winged, winged	geflügelt
brood	Brut
brood (to), to hatch, to incubate	brüten, ausbringen

brooding time, breeding season	Brutzeit
brow tine	Augsprosse
brown bear	Braunbär
browse (to), to graze	äsen
browse (to), to nibble sprouts of young trees and buds	verbeißen
browsing by game, game damage caused by game,game damage	Verbiss, Wildfraß, Wildverbiss,
brush (hair at the end of pizzle of hoofed game), tuft of hair	Pinsel
brush, breech, drag, tail	Fahne, Lunte, Rute, Standarte
bubble, blister, bladder, vesicle	Blase
buck (to)	rammeln
buck shot	Posten (Rehposten)
bucking	Rammelzeit
buff patch, rump patch (hoofed game), target	Spiegel (Schalenwild)
buffalo	Büffel
bugle, hunting horn	Jagdhorn
build (to) an eyrie	horsten
bullet	Geschoss, Kugel
bullet with steel- or copper-jacket	Teilmantelgeschoss
bump (to), to fly against (trees, hunters, fences)	anfliehen
burning scent	warm (Fährte, die Wittrung abgibt)
burr, ring at the base of antler, coronet	Rose (Perlenwulst am unteren Stangenende der Geweihträger)
burrow	Bau
burrow for taking flight, emergency burrow	Fluchtbau, Notbau
burrow, tunnel	Bauröhren von Fuchs und Dachs (Geschleife)
bustard	Trappe

butcher (to), to carve, to cut up	zerwirken
butt (to), to spear, to gore	forkeln
butt, bank of earth, backstop	Kugelfang
butt, owl hut	Hütte, Krähenhütte, Uhuhütte
button spiker, roe buck with knobs instead of horns	Knopfbock
butt plate	Schaftkappe
buttstock	Gewehrkolben, Hinterschaft
buzzard	Bussard, Mäusebussard

C

calf	Kalb
calibre, bore	Kaliber
california condor	Kondor, Kalifornischer
call, cry	Ruf
call (to)	anrufen, mahnen
call (to), to bleat, to whistle	fiepen, blatten
call (to) off a hunter or a dog	abrufen
call (to) off, to announce the end of hunt by signal	abblasen
call (to) on, to blow on	Anblasen des Treibens
caller, roe deer call	Blatter, Rehlocker
call note	Lockruf
call of moufflon, to bleat	meckern
call of rutting hinds	begehren
call on the ground during the coupling season (capercaillie)	Bodenbalz
calling	Lockjagd
calling sound of a kid, cry of fawncalling sound of a kid	Kitzruf
calve (to) (deer), to kid, to lamb, to cub (carnivore), to litter (to farrow, wild boar), to kindle (hare, rabbit)	setzen

camel	Kamel
camouflage, cover, hide	Deckung
campfire	Lagerfeuer
Canada goose	Kanadagans
canine distemper	Hundestaupe
canine tooth of red deer	Grandeln, Eckzahn
canine tooth of the wild sow, hook	Haken (Eckzähne der Bache und des Fuchses)
cant (to)	verkanten
capable of defending	wehrhaft
capercaillie	Auerhahn, Urhahn
capercaillie, wood (great) grouse	Auerwild
capital (stags and roe bucks with big antlers)	kapital
capture (to), to catch	fangen
capture, bag, booty, prey	Beute, Strecke
caracal lynx	Wüstenluchs, Karakal
carcass, kernel	Kern (abgebalgter Raubwild-körper)
carcass of game killed by predaceous animal	Riss
caribou (amerik.), reindeer	Karibu, Rentier
carnassial (canine) tooth	Reißzahn
carrion	Fallwild, Luder
carrion crow	Aaskrähe, Rabenkrähe
carry strap, sling	Gewehrriemen
carted deer	Gatterwild
cartridge	Patrone, Hülse
cartridge bag	Patronenetui
cartridge belt	Patronengürtel
cartridge case	Patronenhülse
carve (to), to butcher, to cut up	zerwirken
case	Futteral
cast (to), to shed the antlers	Abwerfen des Geweihs
cast, pellet, vomits (of predatory birds)	Gewölle
castle of beaver	Biberburg

cat, she-cat	Katze
catch (to), to capture	fangen
catch (to), to capture (predaceous animal)	schlagen
catching of game	Fang
cattle egret	Kuhreiher
cave breeder	Höhlenbrüter
cave, cavern, cavity, sole	Höhle
chain, covey of birds, team of ducks	Kette
chamber, magazine	Patronenlager, Magazin
chamois	Gams, Gamswild, auch Gemse
chamois beard, beard	Gamsbart, Wachler, Bart
chamois buck	Gamsbock
chamois buck with long hairs on the back	Bartgams
chamois doe	Gamsgeiß
chamois horn	Gamskrucken, Krickel, Krucken
chamois kid	Gamskitz
change (to) hair	verfärben
change (to), to move	wechseln
change of dentition	Zahnwechsel
charge (to), to load	laden
charging sound, fighting sound	Sprengruf, Kampfruf
chase (to), to hunt, to pursue, to shoot	jagen, verfolgen
chase, hunting, shooting, stalking	Jagd, Weidwerk
chased, hunted	gejagt
chatter (to), to snarl	keckern
check the zero	Probeschießen
cheek, wild sow	Bache
cheekpiece	Backe
cheeper	Gabelhuhn
cheetah	Gepard, Jagdleopard
chest, breast	Brust
chest hit, shoulder shot	Blattschuss

chilled antlers	Frostgeweih
chital, spotted deer	Axishirsch
choke boring	Chokebohrung, Würgebohrung
cholecyst, gall bladder	Gallenblase
chough	Alpenkrähe
circle (to), to track	kreisen
circle of footprints of a doe and buck on heat	Hexenring
circular beat (hunt), circular drive	Kesseljagd, Kesseltreiben
clava, club, haunch, hind leg, thigh	Keule
claw, clutch, pad, paw	Branke, Brante, Klaue, Pfote, Pranke, Prante
claw, talon	Fang (für Greifvögel), Gewaff, Waffen
clay birding	Tontaubenschießen
clean	stubenrein
cleaning rod	Putzstock
clear (to) out by shooting all game in a shooting-ground	ausschießen
clicking note	knappen, schnalzen, zählen
close (closed) season	Schonzeit
closed footmarks (stag)	Zwang
closed-season rifle	Schonzeitbüchse
cloven-hoofed animal, hoofed game	Paarhufer, Schalenwild
club, clava, haunch, hind leg, thigh	Keule
clutch, claw, pad, paw	Branke, Brante, Klaue, Pfote, Pranke, Prante
coarse – hair	Grannen
coat	Regenmantel
coat, fur, pelt, skin	Balg
coat, hide, leather, skin	Decke (Schalenwild, außer Schwarzwild – Schwarte)
coat, pyles, skin	Schwarte (Dachs)

coati	Nasenbär
coccidiosis	Kokzidiose
cock	Gockel, Hahn
cock, male (for feathered game)	Männchen (Federwild)
cock partridge	Rebhahn
cock pheasant	Fasanenhahn
cock-pigeon	Tauber
cocked	gespannt
cocking lever	Spannhebel
coddes, testicle	Brunftkugel, Kurzwildbret, Geschröte (Hoden von Hunden und Haarraubwild)
cold steel, cold weapon (hunting knife)	blanke (kalte) Waffen
cold trail	kalte Fährte
collared dove	Türkentaube
colonial nester	Koloniebrüter
comb	Kamm
comb (to)	kämmen
combination-barrelled gun	kombinierte Waffe
combination gun	Bockbüchsflinte
common gull	Sturmmöwe
common scoter	Trauerente
commuting game	Wechselwild
conservation area, nature preserve	Naturschutzgebiet
conservation of species, protection	Artenschutz
conservationist	Naturschützer
continuous barking, standing at bay	Standlaut
coot	Blässhuhn
coprophage	Kotfresser
copse,grove,small woods in the landscape for game shelter	Hegebusch, Remise
copulate (to), to court, to display, to pair	balzen

copulate (to), to pair, to mate	paaren
copulation, act of covering,	Beschlag
corkscrew	Korkenzieher
cormorant	Kormoran
coronet, ring at the base of antler, burr	Rose (Perlenwulst am unteren Stangenende der Geweihträger)
coup de graze, finishing shot	Fangschuss
couple	Koppel
court (to), to pair, to copulate, to display	balzen, paaren
courtship, mating, display	Balz
courtship (display) posture	Balzstellung
courtship (mating) dance	Balztanz
courtship (mating) song, courtship call, love song	Balzgesang, Balzarie, Balzlaut, Gesetzl
courtship after sunrise (black grouse)	Sonnenbalz
courtship behaviour	Balzspiel, Balzverhalten
courtship call, love song, courtship (mating) song	Balzlaut, Balzarie, Balzgesang, Gesetzl
courtship plumage	Balzkleid
cover (to) (bear and marmot)	bären
cover (to), to get on	aufsitzen, beschlagen
cover, camouflage, hide	Deckung
covey of birds, chain, team of ducks	Kette
covey, bevy	Gesperre
coyote, praire wolf, barking wolf	Kojote, Präriewolf
coypu, nutria	Biberratte, Nutria, Sumpfbiber
crane	Kranich
cranium, skull	Schädel
craw	Kropf
crawl (to)	schliefen
crest, tuft	Holle
crop (to), to cut a tail, to dock	kupieren
crop, upper breast	Kropf (für Falken)

cross between capercaillie and black grouse	Rackelwild
cross breed (cross-breeding between cat and wild cat)	Blendling (Kreuzung zwischen Haus- und Wildkatze)
cross trace	Schrank, schränken
cross - variety of the red fox	Kreuzfuchs
cross wire	Fadenkreuz
croup, crupper	Kruppe
crow	Krähe
crow-foot	Krähenfuß
crow´s nest	Krähennest
crown	Krone
crupper, croup	Kruppe
crural veins	Brandadern
crutch-pole, perch	Jule
cry, call	Ruf
cry of fawncalling sound of a kid, calling sound of a kid	Kitzruf
cub (to) (carnivore), to calve (deer), to lamb, to kid, to litter (to farrow, wild boar), to kindle (hare, rabbit)	setzen
cull stag, a stag with deformed stags or bad antlers	Abschusshirsch
curl of drake	Erpellocke
curlew	Brachvogel, Großer
curved tail, tail of black-cock	Sichel (gekrümmte Stoßfedern des Birkhahns)
custom	Brauchtum
cut (to) a tail, to dock, to crop	coupieren, kupieren
cut (to) off (forced separation red deer)	absprengen
cut (to) up or to gut the skin of furred game or the plumage of winged game, to blade up	schärfen, aufschärfen
cut (to) up, to butcher, to carve	zerwirken
cutlass, dagger	Hirschfänger

cutting edge	Schneide
cutting tooth, incisor	Schneidezahn

D

dachshund	Dachshund, Dackel, Teckel
dagger, cutlass	Hirschfänger
daily hunting licence	Tagesjagdschein
dall sheep	Dallschaf
damage caused by game, browsing by game,game damage	Verbiss, Wildfraß, Wildverbiss
damage caused by pests, pest-damage	Fraßschaden
dancing ground, display aria	Balzplatz
dazzle (to)	blenden
dead	verendet
decoy	Attrappe
decoy (to),to entice, to lure, to bait,to put food for the game	ködern, ankirren, anludern
decoy bird	Lockvogel
decoy duck	Lockente
decoying a fox, fox calling	Reizjagd, Fuchsreizen
deer call	Hirschruf
deer repellent, antigame protective agent	Verbissschutzmittel
deer with palmed antlers	Schaufler
dehisce (to), to disembowel, to eviscerate, to gut	aufbrechen, ausnehmen, Ausfahren von Waldhühnern
density of game	Wilddichte
dental formula	Zahnformel
dentition, denture, set of teeth	Gebiss
departure for the hunt	Aufbruch zur Jagd
depression (hollow) on the ground during the heat of the fallow buck	Brunftkuhle (Damhirsch)

descend (to) from hind after covering (stag) 1., 2. to fly off (feathered game)	abfallen
detect (to) the details (age, gender, track and so on) of game, to identify	ansprechen
development (formation, growth) of antlers	Geweihbildung
devour (to), to feed	fressen
dew claw, ergot, hind claws	Geäfter, Afterklauen, Oberrücken
diaphragm	Zwerchfell
die (to)	eingehen, verenden
dig (to)	graben
dig (to) up	ausgraben
dinner after hunting	Schüsseltreiben
direction	Folge
discharge (to)	abdrücken
disease, illness, sickness	Krankheit
diseased, ill, sick	krank
disembowel (to), to dehisce, to eviscerate, to gut	aufbrechen, ausnehmen, Ausfahren von Waldhühnern
display (to), to copulate, to court, to pair	balzen
display area, dancing ground	Balzplatz
display in the evening	Abendbalz
display, courtship, mating	Balz
disturb (to)	vergrämen
diurnal	tagaktiv
dive (to) and search for food	gründeln
diving duck	Tauchente
dock (to), to cut a tail, to crop	kupieren, coupieren
dodge (to) about (hare)	Haken schlagen durch den Hasen
doe kid, roe kid, kid, fawn	Geißkitz, Kitz, Rehkitz, Rickenkitz
doe, female (doe) roe, goat	Ricke, Geiß

dog, male of canidae	Rüde
dog breed	Hunderasse
dog breeder	Hundezüchter
dog collar	Halsung
dog distemper	Staupe
dog handler	Hundeführer
dog owner	Hundehalter
dog tape-worm	Hundebandwurm
dog whistle	Hundepfeife
dogging	buschieren
dogging, tracking, to search	absuchen, Suche, Suchjagd
doghouse	Hundehütte
doglead	Hundeleine
dorsal stripe	Aalstreifen, Alstrich,
dorsum, saddle, back	Rücken, Ziemer
double (to)	doppeln
double back	Widergang
double-barrelled gun, double shotgun	Doppelflinte
double crown	Doppelkronel
double pull trigger	Druckpunktabzug
double rifle	Doppelbüchse
double set trigger, hairtrigger	deutscher Stecher, Stecher
double trigger	Doppelabzug
dove, pigeon	Taube
down to fur	hasenrein
down!	down! (nieder!), leg dich!
drag, breech, brush, tail	Fahne, Lunte, Rute, Standarte
drake	Erpel
draw (to) out (uncommon), to pull out, to extract, to remove	aushaken (nicht mehr gebräuchlich), ausnehmen, ausziehen
drink (to)	schöpfen (trinken)
drive (to) small game	abtreiben, durchgehen, durchtreiben, treiben
driven game sights	Drückjagdvisier
driver, beater	Treiber

driving	Standtreiben
driving, beat, battue	Treibjagd
drop	Senkung (Schaftsenkung)
drop (to), to dung	sich lösen
droppings, fewnet, dung	Gestüber, Losung
drum (to)	trommeln
duck	Ente
duiker	Ducker
dull (to) human scent	verwittern
dummy cartridge	Pufferpatrone
dung	Geschmeiß, Losung
dung (to), to drop	lösen (sich)
dung of mating	Balzlosung, Balzpech
dusk	Dämmerung
dwarf-dachshund	Zwergteckel

E

eagle	Adler
eagle owl	Uhu
eagle´s eyrie	Adlerhorst
ear	Lauscher, Luser, Schüssel (Sauen), Teller (Sauen)
ear-ache	Ohrenzwang
ear flap	Ohrschützer
earless seals	Hundsrobben
ears of dog	Behang
ears, hearing, sense of hearing	Gehör
ear tuft, plumicorn, feathered ear	Federohren, Hörner
earth dog	Bauhund, Erdhund
earth hunting, bolting	Baujagd, Erdjagd
earth work	Bauarbeit
eat (to), the dog or the vermin devours a shot game	anschneiden

ecological	ökologisch
eelback, eelstripe, line along back of animals, dorsal stripe	Aalstreifen, Aalstrich,
eider	Eiderente
eight points stag	Achter, Achtender
ejector	Auswerfer, Ejektor
eland	Elenantilope
elephant	Elefant
elk, moose	Elch
elk cow	Elch-Alttier
elk hound	Elchhund
emergency burrow, burrow for taking flight	Notbau, Fluchtbau
empty	leer
enclosed pasture, enclosure, paddock, preserve, reservation	Gehege
end (to) courtship or mating	abbalzen
end (to) rutting	abbrunften
energy of impact	Auftreffenergie
English setter	Lavarack, Englisch Setter
english stock	englischer Schaft
engraving	Gravur
entice (to), to bait, to decoy, to lure,to put food for the game	ködern, ankirren, anludern
entrails, bowels	Aufbruch, Jägerrecht
entrails, intestines	Gescheide
entrance hole, shooting spot	Einschuss
entrance to the burrow of fox or badger,mouth of the burrow	Röhre
equipment	Ausrüstung
ergot, hind claws, dew claw	Geäfter, Afterklauen, Oberrücken
ermine, stoat	Großes Wiesel, Hermelin
escape (to), to flee, to run away, to leap off	fliehen, flüchten, ausbrechen, abspringen
escape, flight	Flucht
European bison	Wisent

European mink	Nerz, Europäischer
even	gerade
eviscerate (to), to dehisce, to disembowel, to gut	aufbrechen, ausnehmen, Ausfahren von Waldhühnern
ewe, sheep	Schaf
excite (to), to irritate, to stimulate	reizen
exit hole	Ausschuss
export licence	Ausfuhrgenehmigung
exterior ballistics	Außenballistik
extract (to), to draw out (uncommon), to pull out, to remove	aushaken, ausnehmen, ausziehen (aushaken nicht mehr gebräuchlich)
extractor	Auszieher
eye	Auge, Licht, Seher
eye wattle, red wattle, red spot above eyes of capercaillie and blackcock,	Flamme, Rose des Auer- und Birkhahnes, Balzrose
eyrie, aerie, hurst, nest	Horst

F

face, sight, vision	Gesicht
falcon	Falke
falcon, varios breeds	Jagdfalke
falconer	Falkner
falconers glove	Falknerhandschuh
falconry, hawking	Beize, Beizjagd, Falkenbeize, Falknerei
fallow-buck of the second head	Knieper
fallow buck of the third head	Löffler
fallow deer	Damwild
fallow doe	Damtier
fallow stag, fallow buck	Damhirsch, Damschaufler
familiar	vertraut

fan, tail	Fächer (Auerhahnstoß)
farrow (to), to litter	frischen
fat of a badger	Dachsfett, Schmalz, Weißes
fat, grease	Feist (Unschlitt)
fat, obese, stout	feist
fawn, doe kid, kid, roe kid	Geißkitz, Kitz, Rehkitz, Rickenkitz
fawn, male calf of red deer, veel	Hirschkalb
fawning season, time for giving birth	Setzzeit, Satzzeit
feathered ear, ear tuft, plumicorn	Federohren, Hörner
feathered game	Federwild
feather follicle	Federbalg
feed (to)	füttern, kröpfen
feed (to) up, to rear	atzen
feed (to), to devour	fressen
feed, feeding, food material	Fraß
feed, food, fodder	Futter, Äsung
feed trough, feed hopper	Schütte
feeding	Fütterung
feeding, feed, food material	Fraß
feeding ground	Äsungsfläche
feeding place	Futterstelle
femal mouflon in the second year	Schmalschaf
female (doe) roe, goat	Geiß, Ricke
female and young (hornless) deer (red, fallow) and elk, hind	Kahlwild
female badger	Dächsin
female calf of red deer, hind calf	Wildkalb
female deer	Mutterwild
female dog, bitch	Hündin
female hare	Häsin, Setzhase
female hogget	Überläuferbache
female horn of a sheep	Stümpfe

female lynx	Luchsin
female of canids (fox), bitch fox, vixen, fuchsin	Bätze, Betze, Fähe, Füchsin
female red deer, adult animal, female adult animal, hind	Alttier, Hirschkuh, Rottier, Tier
fence	Zaun
fence, holding paddock, railing	Gatter, Wildgatter
fennec fox	Wüstenfuchs
ferret	Frettchen
ferret (to)	frettieren
ferreting	Frettierjagd
ferruginous duck	Moorente
fetch (to), to bring, to retrieve	apportieren, bringen
fewnet, droppings, dung	Losung, Gestüber
field: 1. age of hunting dog, 2. land (shotgun)	1. Feld (Altersangabe des Jagdhundes) 2. Feld (gezogener Lauf)
fielfare	Krammetsvogel (Wacholder-drossel)
field for feeding game	Wildacker
field glass, binocular, telescope	Fernrohr, Fernglas, Jagdglas, Prismenglas
field hunting	Feldjagd
fight (to)	kämpfen
fight-tine	Kampfsprosse
fighting sound, charging sound	Kampfruf, Sprengruf
file, arm, tusk	Gewehre (untere Eckzähne des Keilers)
find out, to ascertain, to locate	bestätigen ausmachen
finish (to) off, stab (to), to kill a wounded hoofed game by stabbing with a hanger	abfangen, abnicken, Fang geben
finishing shot, coup de graze	Fangschuss
firearms law	Waffengesetz
firearms licence	Waffenschein
firearms permit	Waffenbesitzkarte
firing lane	Schießbahn

firing-pin, striker	Schlagbolzen
firing position, gun mounting	Anschlag
first aid	Erste Hilfe
first year roe-buck, fawn	Bockkitz
fish-hawk, osprey	Fischadler
flag, thigh	Hose
flank	Flanke
flat	rasant
flatulence, wind	Wind
fledged, ready to fly	flügge
flee (to), to escape, to run away, to leap off	fliehen, flüchten, ausbrechen, abspringen
fleeing, volatile, on the run	flüchtig
flews, lips	Lefzen
flick-knife	Klappmesser
flight, escape	Flucht
flight, group of feathered game	Flug
flight of birds	Strich
flighting of ducks	Entenstrich
flightless	flugunfähig
flinch (to)	mucken (Schuss verreißen)
flock (family of ducks)	Schoof (Schof) für Enten
flushing	Stöberjagd
flushing dog	Stöberhund
fly (to)	streichen, fliegen
fly (to) against (trees, hunters, fences), to bump	anfliehen
fly (to) off	abreiten, abstreichen (Federwild)
fly (to) off (feathered game)	Abstreichen des Federwildes
fly (to) off (feathered game) 1., 2. to descend from hind after covering (stag)	1. Verlassen eines Baumes durch Waldhühner 2. abfallen: nach dem Beschlag (Paarung) vom weiblichen Stück ablassen
fog, mist	Nebel
follow (to) a fleeing animal	mitnehmen

follow (to) a trail or a trace of game (hunting dog)	aufnehmen, anfallen (Annehmen der Fährte durch den Schweißhund)
follow (to) the track of marten in the snow	ausneuen, ausgehen
follow (to) the trail of animals, to track, to trail	spüren, abfährten, abspüren
follow (to) the wounded game (right)	Wildfolge
follow (to) through	mitschwingen
follow (to) with distance	nachhängen
food, feed, fodder	Äsung, Futter
food for the bird of prey	Atzung
food material, feed, feeding	Fraß
foot	Fuß
footmarks on a straight line (fox, lynx, wild cat, wolf)	schnüren
footprint, track	Trittsiegel, Schalenabdruck
foot rot	Moderhinke
forced hunt, hunting with hounds, pack of hunting dogs	Hatz, Hetze
forearm, foreleg	Vorderlauf, Vorderschaft
forefoot, forepaw	Vorderpfote
fore-leg, forearm	Vorderlauf, Vorderschaft
foresight, bead	Korn
fork antlers	Gabelgeweih
fork buck	Gabelbock
fork stag, brocket	Gabelhirsch, Gabler
form, resting place of hare, lair of hare	Sasse
fox	Fuchs
fox calling, decoying a fox	Fuchsreizen, Reizjagd
fox cub	Jungfuchs, Fuchswelpe
foxhound	Foxhound
fox hole	Fuchsbau
fox skin	Fuchsbalg
fox tapeworm	Fuchsbandwurm

fox-terrier	Foxterrier
fray (to), to rub off	fegen
fraying damage	Fegeschaden
free-handed	freihändig
fresh trail	frisch (frische Fährte)
fuants	Dachslosung
fuchsin, female of canids (fox), bitch fox, vixen	Bätze, Betze, Fähe, Füchsin
full choke	Vollchoke
full-jacket bullet	Vollmantelgeschoss
full moon	Vollmond
fur, coat, pelt, skin	Balg
fur, furred game	Haarwild
fur, hair	Haar
fur seal	Seebär
furnish (to)	verecken
furred game	Haarwild
furred predator	Haarraubwild

G

gadwall	Schnatterente
gall bladder, cholecyst	Gallenblase
gallows for partridges, bird sling	Galgen, Hühnergalgen
game	Jagdbeute, Jagdwild
game animals, animals of the chase	jagdbares Wild
game bag	Jagdtasche
game cart	Wildwagen
game cartridge	Jagdpatrone
game crossing, pass, runway	Wildwechsel, Wechsel
game damage	Wildschaden
game damage,damage caused by game,browsing by game	Verbiss, Wildverbiss, Wildfraß

game dealer	Wildhändler
game falls in the fire	Feuer (Wild liegt im Feuer)
game feeding	Wildfütterung, Fütterung
game gun, gun of a hunter	Jagdwaffe
game keeper, game warden	Heger, Jagdaufseher, Jagdhüter
game keeping, wildlife management, game protection	Hege, Wildhege, Wildschutz
game law	Jagdgesetz, Jagdrecht
game-leather, suede	Wildleder
game protection, game keeping, wildlife management	Hege, Wildhege, Wildschutz
game research	Wildforschung
game reserve, wildlife park, wildlife reserve	Wildpark, Wildreservat, Wildschutzgebiet
game science	Wildkunde, Wildtierkunde
game tag	Wildmarke
game tenant, leaseholder	Jagdpächter
gander	Ganter
garganey	Knäkente
gastrolith, gastric calculus	Magensteine
gather (to) under the wing	hudern
gazelle	Gazelle
genealogical table, ancestral register	Ahnentafel
German beagle	Deutsche Bracke
German long-haired pointer	Deutsch Langhaar
German rough-haired pointer	Deutsch Stichelhaar
Gernan short-haired pointer	Deutsch Kurzhaar
German springer	Deutscher Wachtelhund
German terrier	Deutscher Jagdterrier
German wirehaired pointer	Deutsch Drahthaar
gestation period	Tragzeit
get (to) on, to cover	aufsitzen, beschlagen
get (to) out of water (for otter and beaver)	aussteigen
get (to) up	aufstehen
get (to) wind of, to wind, to scent, to smell	winden, Wind bekommen, Wind holen, wittern

gibbon	Gibbon
giraffe	Giraffe
give (to) tongue	verbellen
give (to) tongue (hunting dog), to bark, to yelp, also by foxes during the rutting season	Lautgeben des Jagdhundes, anschlagen, bellen, verbellen, auch Lautäußerung des Fuchses während der Ranz
give (to)tongue in the burrow without contacting to the game	baulaut
gnu, wildebeest	Gnu
go (to) earth or to ground	einfahren (Aufsuchen des Baus), einfallen (Einfahren des Dachses)
go (to) round a thicket (copse) for checking tracks of hoofed game	umschlagen einer Dickung
go (to) with parents-game	führen (Jungwild durch Muttertier führen)
goat, doe, female (doe) roe	Ricke, Geiß
gobble (to)	kullern
going to church	Kirchgang
golden eagle	Königsadler, Steinadler
goldeneye	Schellente
Goldon Retriever	Goldon Retriever
good	brav, gut
good hunting, good luck	Weidmannsheil
goosander	Gänsesäger
goose	Gans
Gordonsetter	Gordonsetter
gore (to), to spear, to butt	forkeln
gorilla	Gorilla
goshawk	Habicht, Hühnerhabicht
gosling	Gössel
grasp, grip, handle, hold	Griff
graze (to), to browse	äsen
grazing shot	Streifschuss
grease, fat	Feist (Unschlitt)
great black-backet gull	Mantelmöwe

great bustard	Großtrappe
great crested grebe	Haubentaucher
great grey shrike	Raubwürger
great Indian rhinoceros	Panzernashorn
great snipe	Doppelschnepfe
greater kudu	Kudu, Großer
grey fox	Grisfuchs
grey heron	Fischreiher, Graureiher
grey seal	Kegelrobbe
grey/white coloured hunting dog with brown patches	Braunschimmel (Brauntiger) häufig bei DD und DK
greyhen	Birkhenne
greylag goose	Graugans
griffon terrier	Griffon
griffon vulture	Gänsegeier
grinder, upper tusk	Haderer
grip, grasp, handle, hold	Griff
grizzly bear	Graubär, Grizzlybär
groove (rifle)	Zug (Büchse)
group of feathered game, flight	Flug
grouse	Raufußhuhn, Schottisches Moorschneehuhn
grove,copse,small woods in the landscape for game shelter	Hegebusch, Remise
grow (to)	schieben
gruntling, young wild boar	Frischling
guide	Jagdführer
gula, gullet, throat	Kehle
gullet, oesophagus, throat	Schlund
gun	Gewehr
gun cover	Gewehrfutteral
gunmaker	Büchsenmacher
gun mounting, firing position	Anschlag
gun of a hunter, game gun	Waffe, Jagdwaffe
gun oil	Waffenöl
gun-shy	schussscheu

gunner, hunter, marksman, markswoman, shooter	Erleger, Jäger, Schütze, Weidmann
gut, intestine, bowels	Darm
gut (to) or to cut up the skin of furred game or the plumage of winged game	aufschärfen, schärfen
gut (to), to disembowel, to dehisce, to eviscerate	ausnehmen, aufbrechen, Ausfahren von Waldhühnern
gut shot, bowel shot	Weidwundschuss
gutter of antler	Riefe (rinnenförmige Vertiefung auf der Oberfläche von Geweihen)
gyr falcon	Gerfalke

H

habitat	Lebensraum
hair, fur	Haar
hair, wool	Wolle
hair balls	Gamskugeln
hair dot	Schürze
hair dot, beard of snipe	Schnepfenbart
hair of a wounded game after shooting	Schnitthaar
hairtrigger, double set trigger	deutscher Stecher, Stecher
half-cock (to)	entspannen
half shovel head	Halbschaufler
hamadryas baboon	Mantelpavian
hammer safety	Schlagstücksicherung
handle, grasp, grip, hold	Griff
hand shy	handscheu
hang (to) together (after the act of covering: dog, fox, wolf)	binden, hängen (Hund, Fuchs, Wolf nach Verpaarung)
hanger	Weidblatt
Hannover bloodhound	Hannoverscher Schweißhund
harbour, lair (hoofed game)	Einstand

harbour seal	Seehund
hard, tough	hart
hare	Hase, Feldhase
hare call	Quäke
hare call, hare pipe	Hasenklage (Hasenquäke)
hare shooting, battue	Hasenjagd
hare´s ear, spoon (fiction)	Hasenohr, Löffel
harp seal	Sattelrobbe
harrier	Weihe
hart, red deer, stag	Hirsch, Rothirsch
hartebeest	Kuhantilope
hatch (to), to brood, to incubate	brüten, ausbringen
hatch (to) out, to hatch, to moult	ausfallen, schlüpfen
hate (to), to swoop	hassen (Stoßen von Greifen, Rabenvögeln und anderen Vögeln auf ihre Feinde)
haunch, clava, club, hind leg, thigh	Keule
hawking bag	Falknertasche
hawking, falconry	Beize, Beizjagd, Falkenbeize, Falknerei
hazel grouse	Haselwild
hazel hen	Haselhuhn
head, antlers, horns	Geweih, Grind, Haupt, Kopf, Gehörn (besser Geweih)
head of game	Stück
head on	spitz
headwind	Gegenwind
healthy: If the hunter has missed a game, it is healthy	gesund
hear (to)	vernehmen
hear (to) the call of stag during the rutting season or of capercaillie and blackcock during the courtship	verhören
hearing, sense of hearing, ears	Gehör

hearing protection	Gehörschutz
hearse, jinnock	Schmaltier (2. Lebensjahr)
heart	Herz
heart shot	Kammerschuss
heat, period of heat	Hitze, Ranzzeit, Ranz, Rollzeit
heat, to be on heat	brunftig
heat period of wild boar, rutting season, brimming time	Rauschzeit
hen	Henne, Rebhenne
hen harrier	Kornweihe
hen pheasant	Fasanenhenne
herd, pack, troop	Rudel
herd of chamois, their kids and bucks	Scharwild, Geraffel (Gamsgeißen, deren Kitze und geringe Böcke)
heron	Reiher
herring gull	Silbermöwe
hibernate (to), to over-winter	überwintern
hibernation, roosting	Winterschlaf
hide (to)	verblenden
hide (to)	drücken: Verstecken des Wildes
hide, coat, leather, skin	Decke (Schalenwild, außer Schwarzwild – Schwarte)
hide, cover, camouflage	Deckung
hide, pyles	Schwarte (Schwarzwild)
hide, screen	Jagdschirm, Schirm
high house	Hochhaus
high seat	Hochsitz, Kanzel
high shot	Hochschuss
hillock	Burgstall
hind, adult animal, female adult animal, female red deer	Alttier, Hirschkuh, Rottier, Tier
hind, female and young (hornless) deer (red, fallow, elk)	Kahlwild
hind calf, female calf of red deer	Wildkalb
hind claws, dew claw, ergot	Geäfter, Afterklauen, Oberrücken

hind leg	Hinterlauf
hind leg, clava, club, haunch, thigh	Keule
hind legs (hare)	Sprünge
hind paw	Hinterpfote
hinged frame	Basküle
hip, loins	Lende
hippopotamus	Nilpferd
hirsute	rauhaarig
hit	Treffer
hit (to) the spine, to stun	krellen, federn
hobby	Baumfalke, Lerchenfalke, Schwalbenfalke
hogget	Überläufer
hold, grasp, grip, handle	Griff
holding paddock, fence, railing	Gatter, Wildgatter
hole breeder	Baumbrüter
holiday sportsman	Sonntagsjäger
hollow point bullet	Hohlspitzgeschoss
honey buzzard	Wespenbussard
hood	Falkenhaube
hooded crow	Nebelkrähe
hoof, cloven hoof	Schale
hoofed game, cloven-hoofed animal	Schalenwild, Paarhufer
hook, canine tooth of the wild sow	Haken (Eckzähne der Bache und des Fuchses)
hook in mount	Einhakmontage
hop (to)	hoppeln
horn of moufflon	Muffelschnecke, Schnecke, Schlauch
horns, antlers, stag´s antlers, horns	Horn, Gehörn (besser: Geweih), Gewichtl oder Gewicht (süddeutsch für Gehörn bzw. Geweih des Rehbocks)
horse-drawn carriage	Jagdwagen
Hungarian short-haired pointer	Ungarisch Kurzhaar, Magyar Vizsla

Hungarian wire-haired pointer	Ungarisch Drahthaar, Magyar Vizsla
hunt (to), to chase	bejagen, verfolgen
hunt (to), to pursue, to chase, to shoot	jagen
hunt (to) with a falcon	beizen
huntable	jagdbar
hunt by sight	sichtlaut
hunted, chased	gejagt
hunter, gunner, marksman, markswoman, shooter	Erleger, Jäger, Schütze, Weidmann
hunters examination	Jägerprüfung
hunters hat, hunters cap	Jagdhut
hunters terms, hunters slang	Weidmannsprache, Jägersprache
hunters yarn	Jägerlatein
hunting, stalking, shooting, chase	Jagd, Weidwerk
hunting accident	Jagdunfall
hunting accident insurance	Jagdunfallversicherung
hunting association	Jagdverband
hunting behaviour	Jagdverhalten
hunting bird	Beizvogel
hunting box, hunting lodge	Jagdhütte, Jagdhaus
hunting control	Jagdschutz
hunting damage	Jagdschaden
hunting dog	Gebrauchshund, Jagdhund
hunting equipment	Jagdausrüstung
hunting ground	Wildbahn
hunting ground (territory)	Jagdgebiet
hunting guest, shooting guest	Jagdgast
hunting horn, bugle	Jagdhorn
hunting instinct	Jagdtrieb
hunting knife, skinner	Jagdmesser, Weidmesser (Jagdnicker, Nicker)
hunting lance, boar spear	Saufeder
hunting lease	Jagdpacht
hunting liability insurance	Jagdhaftpflichtversicherung

hunting licence, hunting permit	Jagderlaubnisschein
hunting methods	Jagdarten
hunting par force	Parforcejagd
hunting party	Jagdgesellschaft
hunting right, shooting right	Jagdrecht
hunting season, shooting season	Jagdzeit, Schusszeit
hunting territory, hunting ground	Jagdrevier
hunting trail, path	Jägersteig, Pirschpfad, Pirschsteig, Pirschweg
hunting with hounds, pack of hunting dogs, forced hunt	Hatz, Hetze
hunting with spaniels	brackieren
hunting year	Jagdjahr
huntsman	Meuteführer, Piqueur
huntsman-like	weidgerecht
hurst, aerie, eyrie, nest	Horst
hut for shooting over decoy owl	Hüttenjagd
hyaena, hyena	Hyäne

I

identify (to), detect (to) the details (age, gender, track and so on) of game	ansprechen
ill, diseased, sick	krank
illness, sickness, disease	Krankheit
illuminated point	Leuchtpunkt
impala	Impala
imperial eagle	Kaiseradler
imply (to), to indicate a shot game	totverweisen
import licence	Einfuhrgenehmigung
imprecise, rough	grob
incisor, cutting tooth	Schneidezahn
incubate (to), to brood, to hatch	brüten, ausbringen
incubation patch	Brutfleck
Indian bear	Lippenbär
Indian elephant	Elefant (Indischer)
indicate (to), to imply a shot game	totverweisen
indigenous	einheimisch
indoor training	Stubendressur
injure (to), to wound	krankschießen, anschießen, anschweißen
insert barrel	Einstecklauf
instinct, natural urge	Naturtrieb
interior ballistics	Innenballistik
internal organs of game: heart, liver, kidneys, lung, spleen	Geräusch
intestine, bowels, gut	Darm
intestines, entrails	Gescheide
introduce (to)	auswildern, einbürgern
invitation for hunting	Jagdeinladung
Irish Setter	Irischer Setter
irregular, abnormal	abnorm

irritate (to), to excite, to stimulate	reizen
ivory	Elfenbein

J

jackal	Schakal
jaguar	Jaguar
jaw, jawbone (lower-, upperjaw)	Kiefer (Unter-u. Oberkiefer)
jay	Eichelhäher, Häher
jess	Geschüh
jinnock, hearse	Schmaltier (2. Lebensjahr)
jog (to), to jolt, to wag	rütteln
jump (to)	überfallen
jump (to) (animals, who are living in burrows), to jump after calling, or bleating (roe buck in the rutting season)	springen (Verlassen des Baues durch Baubewohner), Springen auf´s Blatt (Rehbock)
jump (to) over (capercaillie)	überstellen (Auerhahn wechselt den Ast)
jump trap	Schwanenhals
juvenile, young animal	Junge

K

kangaroo	Känguruh
keen, ready to attack (courageous dog)	scharf sein
kemp, straight hair	Deckhaar
kennel	Hundezwinger, Zwinger
kernel, carcass	Kern (abgebalgter Raubwildkörper)

kestrel	Turmfalke
kettle	Kessel
kickback	Rückstoß
kid (female) of ibex, chamois, deer	Kitzgeiß
kid (to), to calve (deer), to lamb, to cub (carnivore), to litter (wild boar), to kindle (hare, rabbit)	setzen
kid, doe kid, fawn, roe kid	Kitz, Geißkitz, Rehkitz, Rickenkitz
kidney	Niere
kill (to) a wounded feathered game by using a flag feather, pushed into the back of the head (cruelty of animals)	abfedern (Tierquälerei)
kill (to) a wounded hoofed game by stabbing with a hanger, to stab, to finish off	abfangen, abnicken, Fang geben
kill (to) small game after pulling down by hunting dog	abwürgen
kill (to), to bag	erlegen, reißen, töten, strecken, erbeuten
killer, shrike	Würger
killing (shooting), the whole bag of hooofed game	Abschuss
kindle (to) (hare, rabbit), to cub (carnivore), to calve (deer), to lamb, to kid, to litter (to farrow, wild boar)	setzen
kite	Gabelweihe
red kite	Roter Milan
knowledge of tracks and trails	Fährtenkunde

L

Labrador (retriever)	Labrador-Retriever
lachrymal fossa, tear-pit	Tränengrube
ladder	Ansitzleiter, Leiter
lair, harbour	Einstand
lair of hare, resting place of hare, form	Sasse
lamb (to), to calve (deer), to kid, to cub (carnivore), to litter (to farrow, wild boar), to kindle (hare, rabbit)	setzen
lamb of moufflon	Lamm, Schaflamm
lame of a leg	lauflahm
land (shotgun), field (age of hunting dog)	1. Feld (gezogener Lauf), 2. Feld (Altersangabe des Jagdhundes)
landscape management	Landespflege
landscape protection	Landschaftsschutz
lanner falcon	Lannerfalke
lapwing, plover	Kiebitz
large (big) game	Hochwild
Large Münsterländer	Großer Münsterländer
larynx	Drosselknopf, Kehlkopf, Stimmknoten
law of nature	Naturgesetz
lay (to)	legen
lead (to) (hunting dog)	verweisen
lead (to) a shooting-guest	führen (Jagdgast führen)
lead (to), speed-allowance	vorhalten
lead of roe deer	Sprung (Rehwildverband)
leader of a herd, leading animal	Leittier
leading sow	Leitbache
leading stag	Kopfhirsch
leap (to) off, to escape, to run away, to flee	abspringen, fliehen, flüchten, ausbrechen

leaseholder, game tenant	Jagdpächter
leash, strap	Fessel, Langfessel, Riemen
leather, coat, hide, skin	Decke (Schalenwild, außer Schwarzwild – Schwarte)
leave (to) or to change the home district (wood) for searching feed (for hoofed game)	austreten (treten), auswechseln (Schalenwild)
leave (to) the burrow (fox and badger), the lair (hare and wild boar)	ausfahren (Verlassen des Baues, der Sasse, des Lagers von Sauen)
leave (to) the nest by precocial	auslaufen (Nestflüchter verlassen das Nest)
leave (to) the track (hunting dog)	Aufgeben der Fährte durch den Hund
leaving the tree (bird) or the high seat (hunter)	abbaumen
leg (of animal)	Lauf
leg, shank, tarsus	Ständer
legged	geständert
lemming	Lemming
length of the antlers	Stangenlänge
leopard	Leopard
lesser blackbacked gull	Heringsmöwe
lesser kudu	Kudu, Kleiner
lesser spotted eagle	Schreiadler
lesser whitefronted goose	Zwerggans
lesses	Schwarzwildlosung
let (to) loose, uncouple (to), to unleash	schnallen, abhalsen
lever-action rifle	Unterhebelrepetierer
leveret	Junghase, Quarthase (1 Monat), Halbhase (2 Monate), Dreiläufer (halbwüchsig, 3 Monate, ¾ ausgewachsener Hase)
licence	Lizenz

lie (to)	liegen
lie (to) down (dog)	ablegen
lie (to) down, to bed down	niedertun, einschieben
lime-hound	Leithund
line along back of animals, eelback, eelstripe, dorsal stripe	Aalstreifen, Alstrich
line of bore	Seelenachse
line of sight	Visierlinie
lion	Löwe
lion´s mane	Löwenmähne
lioness	Löwin
lips, flews	Lefzen
litter	Geheck
litter (hare and rabbit)	Satz
litter (to) (also to farrow, wild boar), to cub (carnivore), to calve (deer), to lamb, to kid, to kindle (hare, rabbit)	setzen
little, weak, small, of poor condition	gering, schwach
little owl	Steinkauz
live (to) in the burrow, occupied	befahren (bewohnter Bau)
liver	Leber
liver fluke	Leberegel
load (to), to charge	laden
locate (to), to find out, to ascertain	bestätigen, ausmachen
lock	Gewehrschloss, Schloss
loins, hip	Lende
long-eared owl	Waldohreule
long-haired dachshund	Langhaarteckel
long-haired hunting dog	Langhaar
long leash	Schweißriemen
long-tailed duck	Eisente
look at	äugen
loose (to) condition (game) 1., 2. to aim by shooting on the mark	abkommen

loss (the) of a branch of antlers by a stag through fighting a rival	abkämpfen (Stangen oder Geweih im Kampf abbrechen)
love song, courtship call, courtship (mating) song	Balzarie, Balzgesang, Balzlaut, Gesetzl
lower breast	Stich
lower jaw, mandible	Unterkiefer
low game shooting	Niederjagd
low house	Niederhaus
lubricate (to)	Waffe einfetten
lung	Lunge
lung worm	Lungenwurm
lure	Federspiel
lure (to),to decoy, to bait, to entice,to put food for the game	ködern, ankirren, anludern
lure, bait	Köder, Lockmittel
lynx	Katzenluchs, Luchs
lyre shaped tail, tail, playing tail (blackcock)	Leier, Schar, Schere, Spiel (Birkhahn)

M

magazine, chamber	Magazin, Patronenlager
magpie	Elster
maize	Mais
make (to) a driving for small game	abtreiben, durchgehen, durchtreiben, treiben
male (for feathered game), cock	Männchen (Federwild)
male calf of red deer, fawn, veel,	Hirschkalb
male cat	Kuder
male hare	Rammler
male kid	Bockkitz
male of canidae, dog	Rüde
mallard	Stockente
mammal	Säuger

mammary glands of animals, udder	Gesäuge, Spinne
mammoth	Mammut
mandible, lower jaw	Unterkiefer
mane, rutting mane	Mähne, Brunftmähne, Kragen
maned wolf	Mähnenwolf
mange, scab	Räude
mannubrium, trunk	Rüssel
mantle, pallium	Mantel
maral	Maral
Marco Polo argali	Marco-Polo-Schaf
mark	Zeichen
mark (to)	1. markieren, 2. zeichnen
mark (to) the killed hoofed game with a branch	verbrechen
mark, sign, spoor, track	Pirschzeichen, Pürschzeichen
marking of territory	Reviermarkierung
marksman,markswoman (huntress),gunner, hunter, shooter	Erleger, Jäger(in), Schütze, Weidmann
marmot	Murmeltier
marsh harrier	Rohrweihe
marten	Marder
mask	Maske, Zügel
mast	Mast
master	Jagdleiter (Eigentümer der Meute)
mate (to), to copulate, to pair	paaren, reihen
mating (courtship) dance	Balztanz
mating flight	Balzflug
mating season	Paarungszeit
mating season of ducks	Reihzeit
mating, display, courtship	Balz
mature, ripe	reif
maxilla, upper jaw	Oberkiefer
maxillary branch, ramus mandibuli	Kieferast

meet	Sammelplatz
meeting point	Treffpunkt (Sammelplatz)
merlin	Merlin
migrate (to)	ziehen (Zugvögel)
migration	Vogelzug
migratory bird	Zugvogel
migratory bird, bird of passage	Zugvogel, Strichvogel
milk dentition, primary dentition,	Milchgebiss
mirror (duck)	Spiegel (Ente)
misfire	Versager
miss	Fehlschuss
miss (to)	fehlen
mist, fog	Nebel
molar, molar tooth, back tooth	Molar, Backenzahn
monk seal	Mönchsrobbe
monkey	Affe
monogammous	monogam
monogamy	Monogamie
Montagu´s harrier	Wiesenweihe
moon (old m. = abnehmend, waxing m. = zunehmend)	Mond
moose, elk	Elch
mosquito net	Mückenschleier
mosquito repellent	Mückenschutzmittel
moufflon	Muffelwild, Mufflon
moult (for wildfowl)	Raue (Mauser beim Wasserfederwild)
moult (to), moulting, if a female dog on heat bleeds from vagina	färben (Haarwechsel und Schweißen der läufigen Hündin aus der Schnalle)
moult (to)	mausern, schildern
moult drake	Mausererpel, Rauerpel
moult, moulting	Haarwechsel, Mauser, Federwechsel, Raue
moult (to), to hatch out, to hatch	ausfallen, schlüpfen

mount (to) the trophies on a board, to attach	aufsetzen: Trophäe auf Holzbrett montieren
mountain goat	Steinbock
mountain hare, blue hare	Schneehase
mountain lion, puma	Puma
mouse call, mouse pipe	Mauspfeifchen
mouth	Äser, Geäse
mouth of the burrow,entrance to the burrow of fox or badger	Röhre
mouth, muzzle	Fang, Maul
move (to), to change	wechseln
move (to), to slide	rutschen
move (to) slowly (hoofed game)	ziehen (Schalenwild)
movement of the wounded game after shooting, to mark	zeichnen
muntjac	Muntjak
murderer	Mörder
muscular shot	Wildbretschuss
musk	Bisam
musk beetle	Moschusbock
musk deer	Moschushirsch
mutes	Greifvogellosung
mute swan	Höckerschwan
muzzle	Fang, Laufmündung, Maulkorb
muzzle blast	Mündungsfeuer
muzzle cover	Mündungsschoner
muzzle, mouth	Fang, Maul
myxomatosis	Kaninchenpest, Myxomatose

N

nail (to)	nageln
nape, back of the neck, nucha	Genick
nasal bone	Nasenbein
natural urge, instinct	Naturtrieb
nature conservation	Naturschutz
nature preserve, conservation area	Naturschutzgebiet
neck (hoofed game)	Träger (Hals des Schalenwildes außer Schwarzwild), Hals (Niederwild)
neckring, ring (cock pheasant)	Ring
nest	Nest
nest, aerie, eyrie, hurst	Horst
nest (to)	nisten
nest (clutch) of eggs	Gelege (Brut)
nesting	Nestbau
nesting box	Nistkasten
nesting hole	Nisthöhle
nesting place, breeding place	Brutplatz, Nistplatz
nestling	Nestling
nestling, nidicolous bird	Nesthocker
nibble (to) at sprouts of young trees and buds, to browse	verbeißen
Night-glass	Nachtglas
nocturnal	nachtaktiv
noisy, babbling	weidlaut
nose	Nase, Windfang
nucha, back of the neck, nape	Genick
nuptial plumage, nuptial coloration	Hochzeitskleid
nurse (to), to suckle, to breast-feed	säugen
nutcracker	Tannenhäher
nutria, coypu	Biberratte, Nutria, Sumpfbiber

O

obedience	Appell, Gehorsam
obese, fat, stout	feist
objective lens	Objektiv
obligate runway	Zwangswechsel
occupied, to live in the burrow	befahren (bewohnter Bau)
ocelot	Ozelot
ocular lens	Okular
odd, uneven	ungerade
oesophagus, gullet, throat	Schlund
old, aged	alt
old solitary, big wild boar, singular	grobes Schwein, Hauptschwein
omasum, third stomach of ruminants, psalterium	Blättermagen
omnivore	Allesfresser
on heat, ruttish	läufig
on the run, fleeing, volatile	flüchtig
open sight	offene Visierung (Kimme und Korn)
optimum sighting in range	günstigste Einschießentfernung (GEE)
orang-utan	Orang-Utan
oryx gazella	Oryxantilope
osprey, fish-hawk	Fischadler
ostrich	Strauß
otter	Fischotter
otter (otter dog)	Otter (männlicher Otter)
outfitter	Ausrüster
over-under rifle	Bockdoppelbüchse
over-under shotgun	Bockdoppelflinte
overpreserve (to)	überhegen
overreaching	Beitritt
over-winter (to), to hibermate	überwintern
owl	Eule, Kauz
owl hut, butt	Hütte, Krähenhütte, Uhuhütte
owner of shooting grounds	Jagdherr (gilt auch für Pächter)

P

pack (of wild boars)	Rotte
pack of hounds	Hundemeute, Meute, Saumeute
pack of hunting dogs, hunting with hounds, forced hunt	Hatz, Hetze
pack, herd, troop	Rudel
pad, claw, clutch, paw	Branke, Brante, Klaue, Pfote, Pranke, Prante
paddock,enclosed pasture, enclosure, preserve, reservation	Gehege
pair	Paar
pair (to), to copulate, to court, to display	balzen, paaren
pallium, mantle	Mantel
palmate antlers, palm	Schaufelgeweih
palmate foot, webbed foot	Ruder, Latschen
palmed head	Schaufelkrone
panther	Panther
parasite	Parasit
Parson-Jack-Russell-Terrier	Parson-Jack-Russell-Terrier
partial migrant	Teilzieher
partridge	Rebhuhn
partridge shooting	Hühnerjagd
pass, game crossing, runway	Wechsel, Wildwechsel
passage in the evening	Abendstrich
passage of snipes	Schnepfenstrich
path	Pass
path, hunting trail	Jägersteig, Pirschpfad, Pirschsteig, Pirschweg
path of the target	Flugrichtung
paunch, rumen, stomach	Pansen, Weidsack
paunch content	Panseninhalt
paw, claw, clutch, pad	Branke, Brante, Klaue, Pfote, Pranke, Prante
paw (to), scraping with the fore legs	plätzen

pea rifle	Tesching
pearls	Perlen
pedicle, antler base	Rosenstock
pedrigree	Stammbaum
peel (to)	schälen
pellet, cast, vomits (of predatory birds)	Gewölle
pelt, coat, fur, skin	Balg
pelvic joint (cloven hoofed game) 1., 2. lock (gun)	Schloss: 1.Beckenknochen-verbindung, 2.Gewehrschloss
pendulum beam	Pendelstange
penis, pizzle (predaceous animals)	Brunftrute, Feuchtglied, Fruchtglied (Hund, Haarraubwild), Rute (Hund, Haarraubwild)
perch (to)	aufbaumen
perch (to), to settle	einfallen (Federwild lässt sich nieder)
perch, crutch-pole	Jule
peregrine falcon	Wanderfalke
period of heat, heat	Hitze, Ranz, Ranzzeit, Rollzeit
permanent dentition	Dauergebiss
persue (to)	nachprellen
pest, vermin	Raubzeug
pest, zoonosis	Tierseuche
pest-damage, damage caused by pests	Fraßschaden
pharynx	Rachen
pheasant	Fasan, Jagdfasan
pierce (to), to search for food (snipe), to sting, to worm	stechen, wurmen (Schnepfe)
pigeon	Taube
piglet	Frosch (Frischling bis Alter von 4 Wochen)
pinch	knautschen
pin feather	Malerfeder, Schnepfengrandel
pine marten	Baummarder, Edelmarder, Tannenmarder

pinion, wing	Schwinge, Flügel
pinkfooted goose	Kurzschnabelgans
pintail	Spießente
piss (to), to urinate	feuchten, nässen
pistol grip	Pistolengriff
pizzle (predaceous animals), penis	Brunftrute, Feuchtglied, Fruchtglied (Hund, Haarraubwild), Rute (Hund, Haarraubwild)
place of carrion, bait place	Luderplatz
place roe buck	Platzbock
place stag	Platzhirsch
plate trap	Tellereisen
playing tail, lyre shaped tail, tail (blackcock)	Leier, Schar, Schere, Spiel (Birkhahn)
plover, lapwing	Kiebitz
pluck	Rupfung
plumage	Federkleid, Gefieder
plumicorn, feathered ear, ear tuft	Federohren, Hörner
poach (to)	wildern
poacher	Wilddieb, Wilderer
poaching	Wilderei
pochard	Tafelente
pocket of gland, venturi tube	Saugloch, Schmalzröhre, Stinkloch
point, branch, prong of antler, tine	Ende, Geweihzacken, Sprosse
point (to), hunting dog is standing in front of low game	vorstehen
point blank shot	Fleckschuss
pointer	Hühnerhund, Pointer
pointing dog	Vorstehhund
point of aim	Haltepunkt
point of impact	Treffpunkt (Auftreffpunkt)
point (to) pronely	vorliegen (Bauhund im Fuchsbau)
polar bear	Eisbär
polecat	Waldiltis

polygamous	polygam
polygamy	Vielehe
poodle pointer	Pudelpointer
population	Population
porcupine	Stachelschwein
portable firearm	Handfeuerwaffe
post (to) the guns	besetzen
post (to) the hunters and beaters around the hunting district, to surround	abstellen, umstellen
postmating	Nachbalz
pothunter	Aasjäger, Schießer
powder	Pulverladung
praire wolf, barking wolf, coyote	Kojote, Präriewolf
prayer	Gebet
precocial bird	Nestflüchter
predaceous animals and birds preying on game	Raubwild
predator	Räuber
predatory (raptorial) bird, bird of prey, birds preying on game	Greifvogel, Raubvogel
pregnant, to be pregnant (bitch)	Trächtigwerden von Raubwild-fähe und Hündin, dickgehen, hochbeschlagen, innehaben
prehensile toe	Greifzehe
preparator	Präparator
preserve (to) 1., 2. game with wounded legs that doesn´t put all foots on the ground	schonen: 1. hegen, 2. Nichtaufsetzen aller Läufe kranken Schalenwildes nach Laufverletzung
preserve (to), to protect	hegen
preserve, enclosed pasture, enclosure,paddock, reservation	Gehege
prevent (to), game damage	Wildschadensverhütung
prey, bag, booty, capture	Beute, Strecke
prick (to), to set the trigger	einstechen, stechen

pricket, spiker, stag with first antlers	Spießer (Schmalspießer)
primary dentition, milk dentition	Milchgebiss
primer	Zündhütchen
productive (eatable) game	Nutzwild
professional hunter	Berufsjäger
pronghorn	Gabelbock (Antilocapra americana)
prong of antler, branch, point, tine	Ende, Geweihzacken, Sprosse
protect (to), to preserve	hegen, überhegen
protected landscape	Landschaftsschutzgebiet
protection, conservation (preservation) of species	Artenschutz
psalterium, third stomach of ruminants, omasum	Blättermagen
ptarmigan	Alpenschneehuhn, Schneehuhn
puff (to) 1. wild boar (to blow), 2. hunting horn	blasen. 1. Schwarzwild, 2. Jagdhorn
pull (to)	ziehen (herausziehen)
pull (to) down	niederziehen
pull (to) out, to draw out (uncommon), to extract, to remove	aushaken (nicht mehr gebräuchlich), ausnehmen, ausziehen
pull (to) the trigger	abdrücken
puma, mountain lion	Puma
punch (to) (hare, rabbit) 1., 2. bay (to) the dogs by stags or wild boars. 3. to beat off a rival by stag or wild boar.	abschlagen
punish (to)	strafen
pup, whelp	Welpe
purse net	Frettiernetz, Kaninchenhaube, Netz
pursue (to), to chase, to hunt, to shoot	jagen

put (to) food for the game, to lure with a bait (carcass), to bait, to decoy, to entice, to lure	ankirren, anludern, ködern
put (to) up, to rouse	hochmachen
put (to) up the trap 1., 2. to bay (for big game)	stellen: 1. Stellen einer Falle, 2. Wild stellen
putrefy (to), to rot	verludern
pyles, coat, skin	Schwarte (Dachs)

Q

quail	Wachtel

R

rabbit	Kaninchen, Wildkaninchen
rabbit burrow (hole)	Kaninchenbau
rabbit punch	Genickfang
rabies	Tollwut
raccoon	Waschbär
raccoon dog	Marderhund
railing, holding paddock, fence	Gatter, Wildgatter
rails	Rallen
ram	Widder
ram lamb	Widderlamm
ram´s horns	Widdergehörn (-geweih für Rehbock, abnorm)
ramus mandibulli, maxillary branch	Kieferast
rat	Ratte
rattle	Klapper
rattle (to)	klappern
raven	Kolkrabe

ready for catching	fängisch
ready position	Jagdanschlag
ready to attack (courageous dog), keen	scharf sein
ready to fly, fledged	flügge
rear (to), to feed up	atzen
rear sight notch, sighting notch	Kimme
receiver	Verschlussgehäuse
recent snow	Neue, Neuschnee, Spurschnee
rectrix, rudder feathers	Steuerfedern
red-breasted merganser	Mittelsäger
red-crested pochard	Kolbenente
Red Data Book	Rote Liste
red deer, hart, stag	Edelwild, Rothirsch, Rotwild
red deer calf	Rotwildkalb
red fox	Rotfuchs
red grouse	Schottisches Moorschneehuhn
red kite	Roter Milan
red-legged partridge	Rothuhn
red spot above eyes of capercaillie, blackcock and other winged game, eye wattle, red wattle	Flamme, Rose des Auer- und Birkhahnes sowie sonstigen Federwildes, Balzrose
red track	Rotfährte
red track, blood trail	Schweißfährte, Wundfährte
red wattle, red spot above eyes of capercaillie, blackcock and other winged game, eye wattle	Flamme, Rose des Auer- und Birkhahnes sowie sonstigen Federwildes, Balzrose
reindeer, caribou (amerik.)	Rentier, Karibu
release (to)	aussetzen, auswildern
reload (to)	nachladen, wiederladen
remove (to), to draw out (uncommon), to extract, to pull out	aushaken (nicht mehr gebräuchlich), ausnehmen, ausziehen
repeat (to)	repetieren
repeater, repeating rifle	Repetierbüchse
reservation,enclosed pasture, enclosure, paddock,preserve	Gehege

resident game	Standwild
resting place of hare, lair of hare, form	Sasse
resting plumage, winter plumage	Winterkleid
retrieve (to), to bring, to fetch	apportieren, bringen
retriever	Apportierhund
revolver	Revolver
reward (to)	genossen machen
rhinoceros	Nashorn
rib	Laufschiene
rib	Rippe
ricochet hit	Querschläger
rifle	Büchse
rifle cartridge	Büchsenpatrone
rifled	gezogen
rifled barrel	Büchsenlauf
rifling	Drall
ring (to)	beringen
ring, neckring (cock pheasant)	Ring
ring at the base of antler, coronet, burr	Rose (Perlenwulst am unteren Stangenende der Geweihträger)
ringed seal	Ringelrobbe
ripe, mature	reif
rise (to)	hochwerden
roan	Pferdeantilope
roar (to), to bell	melden, orgeln, röhren, schreien
roaring stag	Brunfthirsch
rock partridge	Steinhuhn
Rocky Mountain goat	Schneeziege
rodent	Nager
roe	Reh
roe buck	Bock (Rehbock)
roe buck with knobs instead of horns, button spiker	Knopfbock
roe deer	Rehwild
roe deer call, caller	Blatter, Rehlocker

roe deer calling	Blattjagd
roe doe (in the second year)	Schmalreh
roe kid, doe kid, fawn, kid	Geißkitz, Kitz, Rehkitz, Rickenkitz
rook	Saatkrähe
roosting, hibernation	Winterschlaf
roosting tree	Schlafbaum
root to	brechen
rooting place	Gebräch
rostrum	Schnabel
rot (to),to putrefy	verludern
rotten game, to rot	verhitzen
rough, imprecise	grob
rough-haired dachshund	Rauhaarteckel
rough-haired pointer	Stichelhaar
rough-legged buzzard	Raufußbussard
roughage	Raufutter
rouse (to)	Wild aus dem Treiben drücken
rouse (to), to put up	hochmachen
rouse (to) wounded game	aufmüden
rousing game	Drückjagd
rub (to) off, to fray	fegen
rubber boots	Gummistiefel
rucksack, backpack	Rucksack
rudder feathers, rectrix	Steuerfedern
ruddy shelduck	Rostgans
ruff	Kampfläufer
rumen, paunch, stomach	Pansen, Weidsack
rumen content	Panseninhalt
ruminant	Wiederkäuer
ruminate (to)	wiederkäuen
rummage (to)	stöbern
rump patch (hoofed game), buff patch, target	Spiegel (Schalenwild)
run (to)	rennen
run (to) away, to escape, to flee, to leap off	fliehen, flüchten, ausbrechen, abspringen
running shot	Schuss auf flüchtendes Wild

runway, game crossing, pass	Wildwechsel, Wechsel
rut (to)	ranzen
rut (to), to brim	rauschen
rut of deer	Hirschbrunft
rutting-mane, mane	Brunftmähne, Mähne, Kragen
rutting mark, rutting spot	Brunftfleck
rutting place	Brunftplatz
rutting season	Brunft, Brunst
rutting season of roe	Blattzeit
rutting season, period of heat, brimming time	Rauschzeit
rutting smell	Brunftgeruch
ruttish, on heat	läufig

S

sable	Zobel
sable antelope	Rappenantilope
sacrilege against hunting, unlawful shooting	Jagdfrevel, Jagdvergehen
saddle, back, dorsum	Ziemer, Rücken
saddlecloth, white back of certain mammals (ram)	Sattel, Schabracke, Schneefleck
safari	Safari
safety	Sicherung, Sicherheit
salt lick	Salzlecke, Sulze, Leckstein
scab, mange	Räude
scald (to), to boil	abkochen
scan (to)	abglasen, ableuchten
scattergun, break-down gun	Kipplaufwaffe
scent (to), to wind, to smell, to get wind of	wittern
scent, scented bait	Witterung (Jägersprache: Wittrung)
scent-bag (hunting dogs)	Schleppe (Geschleife)

scent gland	Brunftfeige, Duftdrüse, Nelke, Viole
scented trail, trail drag, scent bag	Schleppe, Geschleife (künstliche Wundfährte für den Hund)
scimitar- horned oryx	Säbelantilope
scope mount	Zielfernrohrmontage
scraping with the fore legs, to paw	plätzen
screen, hide	Jagdschirm, Schirm
scut, (tail for wild boar)	Wedel (Pürzel für Schwarzwild)
sea otter	Seeotter
seam	Naht
sear safety	Stangensicherung
search (to) for food (snipe), to pierce, to sting, to worm	stechen, wurmen (Schnepfe)
search (to) with a hunting dog	revieren
search (to), tracking, dogging	absuchen, Suche, Suchjagd
seat stick, shooting seat, stick	Jagdstuhl, Sitzstock
secondary feather	Armschwinge
second stag	Beihirsch
see (to)	eräugen, wahrnehmen
seed tree	Mastbaum
seek (to) wounded game and to retrieve it	verlorenbringen
seize (to)	packen
selector	Umschaltschieber
sense of hearing, ears, hearing	Gehör
sense of vision	Gesichtssinn
set at safe	sichern (Waffe)
set back (to)	zurücksetzen (geringeres Geweih als im Vorjahr)
set of teeth, dentition, denture	Gebiss
set (to) the trigger, to prick	stechen, einstechen
setter	Setter
settle (to), to perch	einfallen (Federwild lässt sich nieder)
shaft, stock	Schaft

shag	Krähenscharbe
shank, leg, tarsus	Ständer
sheath	Scheide
she-cat, cat	Katze
shed (to) the antlers, to cast	Abwerfen des Geweihs
shed antler	Abwurfstange
shedding of antlers	Abwurf
sheep, ewe	Schaf
shelduck	Brandente
shield	Panzer
shield (capercaillie, hazel cock, cock pheasant, partridge) 1., 2. blade (wild boar), shoulder blade (wild boar), shoulder	1. Schild (Auerhahn, Haselhahn, Fasan, Rebhuhn), 2. Blatt und Schulterblatt (Schwarzwild)
shoot (to), to chase, to hunt, to pursue	jagen
shoot (to) at game	anbringen, antragen, beschießen, schießen
shoot (to) off	abschießen
shooter, hunter, gunner, marks(wo)man	Jäger(in), Erleger, Schütze, Weidmann
shooting, chase, hunting, stalking	Jagd, Weidwerk
shooting (killing), the whole bag of hooofed game	Abschuss
shooting driven game	Vorstehtreiben
shooting-guest, hunting guest	Jagdgast
shooting in the mating season	Balzjagd
shooting licence	Jagdschein
shooting light	Büchsenlicht
shooting of low game	Niederwildjagd
shooting plan	Abschussplan
shooting practice	jagdliches Schießen
shooting range	Schießstand
shooting right, hunting right	Jagdrecht
shooting season, hunting season	Jagdzeit, Schusszeit

shooting seat	Drückjagdbock
shooting seat, stick, seat stick	Sitzstock, Jagdstuhl
shooting spot, entrance hole	Einschuss
shooting station	Stand
shore birds	Limikolen
short bellow	knören, trensen
short-eared owl	Sumpfohreule
short rifle	Stutzen
short-toed eagle	Schlangenadler
shot charge	Schrotladung
shot column	Schrotgarbe
shot distance	Schussentfernung
shotgun	Einzellader, Flinte
shotgun barrel	Flintenlauf
shotgun cartridge	Schrotpatrone
shotgun shooter	Flintenschütze
shot mark	Schusszeichen
shot of rifle	Büchsenschuss
shot-proof	schussfest
shoulder	Blatt, Schulter
shoulder shot, chest hit	Blattschuss
shoulder, blade, shoulder blade (wild boar) 1., 2. shield (capercaillie, hazel cock, cock pheasant, partridge)	1. Blatt und Schulterblatt (Schwarzwild). 2. Schild (Auerhahn, Haselhahn, Fasan, Rebhuhn)
shoveler	Löffelente
shrike, killer	Würger
sick, ill, diseased	krank
sickness, disease, illness	Krankheit
side wind	halber Wind (Seitenwind)
sight	Visier
sight, face, vision	Gesicht
sighting notch, rear sight notch	Kimme
sign, mark, spoor, track	Pirschzeichen, Pürschzeichen
signals blown by hunters during the driving	Jagdsignal

sika deer	Sikahirsch
silent (to be)	verschweigen
silver fox	Silberfuchs
single set trigger	Rückstecher
single shot	einläufige Flinte
single stage trigger	Flintenabzug
single trigger	Einabzug
singular, big wild boar, old solitary	grobes Schwein, Hauptschwein
sit (to) up for shooting	ansetzen, ansitzen
sitting, stand hunting	Ansitz, Anstand
six points stag	Sechser, Sechsender
skeet shooting	Skeetschießen
skin (to)	abbalgen, abschwarten, abstreifen, streifen
skin, coat, fur, pelt	Balg
skin, coat, pyles	Schwarte (Dachs)
skin, hide, coat, leather	Decke (Schalenwild, außer Schwarzwild – Schwarte)
skinner, hunting knife	Jagdmesser, Weidmesser (Jagdnicker, Nicker)
skin of stag	Hirschdecke
skittle	Kegel (Männchen machen)
skull, cranium	Schädel
sleeping (roost) place	Schlafplatz
slide (to), to move	rutschen
sling, carry strap	Gewehrriemen
slip (to) in	einschliefen
slot	Eingriff (Trittsiegel)
slug	Flintenlaufgeschoss
small. of poor condition, weak, little	gering, schwach
small bore, small caliber	Kleinkaliber
small (low) game	Niederwild
small game hunter	Niederwildjäger
Small Münsterländer	Kleiner Münsterländer
small shot	Schrot, Hagel

smell (to)	bewinden, riechen
smell (to), to scent, to wind, to get wind of	wittern
smew	Zwergsäger
smokeless	rauchlos
snap shooting	abpassen
snare	Schlinge, Falle
snarl (to)	murren
snarl (to), to chatter	keckern
snipe	Schnepfe, Bekassine, Himmelsziege
snipe, which is staying the whole year (also in winter) in the home location	Lagerschnepfe
snout	Gebrech
snout	Scheibe, Wurf (Windfang)
snow goose	Schneegans
snow leopard	Schneeleopard
social courtship	Gruppenbalz
softpoint round nose bullet	Teilmantelrundkopf-Geschoss
sole, cave, cavern, cavity	Höhle
somersault (to)	rollieren, sich überschlagen
sound (to) death	verblasen
sow, wild pig	Sau
span, spread	Auslage
spaniel	1. Bracke. 2. Wachtelhund. 3. Spaniel
sparrow hawk	Sperber
sparrow hawk (male)	Sprinz
spayed animal, sterile hind	Gelttier
speaking while trailing	spurlaut
spear (to), to gore, to butt	forkeln
speed-allowance, to lead	vorhalten
spiker, pricket, stag with first antlers	Spießer (Schmalspießer)
spike stag, spiker	Spießbock
spleen	Milz

splinter of bone, bone fragment	Knochensplitter
spoon (fiction), hare´s ear	Löffel, Hasenohr
spoor, mark, sign, track	Pirschzeichen, Pürschzeichen
spoor, track, trail	Fährte, Spur, Wildspur
sport (to), (stags play with their antlers)	scherzen (Spielen der Hirsche mit ihrem Geweih, Kraft erproben)
spotted crake	Tüpfelsumpfhuhn
spotted deer, chital	Axishirsch
spotted eagle	Schelladler
spotted hyena	Tüpfelhyäne
spread (to)	streuen (Munition)
spread cone for a shotgun	Zusammenhalten der Schrote bei nicht streuender Flinte
spread, span	Auslage
springbuck, springbok	Springbock
springer spaniel	Springer, Spaniel
spur (pheasant)	Sporen, Sporn
St. Hubert´s day	Hubertustag
stab (to), to kill a wounded hoofed game by stabbing with a hanger, to finish off	abfangen, abnicken, Fang geben
stag, hart, red deer	Hirsch, Rothirsch
staghound	Hirschhund
stag in grease	Feisthirsch
stag of ten points	Zehnender, Zehner
stag of twelve points	Zwölfender, Zwölfer
stag´s antlers, horns	Hirschgeweih
stag with cruciform antlers	Kreuzbock
stag with deformed stags, cull stags or bad antlers	Abschusshirsch
stag with first antlers, pricket, spiker	Schmalspießer
stag with velvet-covered antlers	Kolbenhirsch
stalk (to)	pirschen, pürschen
stalk (to) a calling capercaillie by bounds	anspringen

stalk (to) game	anpirschen
stalker	Pirschjäger
stalking	Birsch, Pirsch, Pürsch
stalking, chase, hunting, shooting	Jagd, Weidwerk
stand (to) at gaze, to stop, to wind	verhoffen (sichern)
stand hunting, sitting	Anstand, Ansitz
standing at bay, continuous barking	Standlaut
stay (to) in a certain district (game) 1., 2. to stay (female deer) and want to be covered by a stag, 3. to point (the hunting dog signals the presence of low game)	stehen: 1. Einstand, 2. weibliches Schalenwild steht dem männlichen Stück beim Beschlag Beschlag. 3. vorstehen
stay (to) in underwood (wild boars)	stecken (Sauen)
steel trap	Abzugseisen
step	Schritt
sterile hind, spayed animal	Gelttier
stick, seat stick, shooting stick	Jagdstuhl, Sitzstock
stick (to) (wild boars stay in underwood)	stecken (Sauen)
stick (to) to the track	halten (Fährte halten durch den Jagdhund)
still hunt at dawn	Morgenansitz
still hunt at dusk	Abendansitz
stimulate (to), to excite, to irritate	reizen
sting (to), to search for food (snipe), to pierce, to worm	stechen, wurmen (Schnepfe)
stoat, ermine	Großes Wiesel, Hermelin
stock, shaft	Schaft
stock dove	Hohltaube
stock of game	Bestand, Wildbestand
stock of low game	Besatz (gilt nur für Niederwild), siehe auch Wildbestand

stocking rate	Besatzdichte
stomach, belly	Magen, Bauch
stomach, paunch, rumen	Pansen, Weidsack
stop	Stop (Stirnabsatz)
stop (to), to wind, to stand at gaze	verhoffen (sichern)
stork	Storch
stout, fat, obese	feist
straight hair, kemp	Deckhaar
strap, leash	Riemen
stride	Schrittlänge
strike (to)	hauen
striker	Schlagbolzen
striking at the dogs with the tusks by wild boars	schlagen (Verletzen der Verfolger von Sauen durch ihr Gewaff)
strip (to) the skin	abschärfen
stripping damage	Schälschaden
strong	stark
study of shooting	Schießlehre
stuff (to)	präparieren
stun (to), to hit the spine	federn, krellen
sub terminal band	Endbinde, Stoßendbinde
suckle (to) to breast-feed, to nurse	säugen
suede, game-leather	Wildleder
support (to) the riffle against a tree	anstreichen
surround (to), post (to) the hunters and beaters around the hunting district	abstellen, umstellen
sur-royal, a branch under the crown of stag´s antlers	Wolfssprosse (tiefangesetztes Kronenende des Rothirsch-geweihs)
swan	Schwan
swim (to)	rinnen
swim (to) across	durchrinnen
swine fever	Schweinepest

swoop (to), to hate	hassen (Stoßen von Greifen, Rabenvögeln und anderen Vögeln auf ihre Feinde)

T

tag	Luntenspitze
tail (cock pheasant)	Spiel, Stoß (Fasanenhahn)
tail (wild boar and falcon)	1. Bürzel, Pürzel, Federlein (Schwarzwild). 2. Schwanz (für Falken)
tail of black-cock, curved tail	Sichel (gekrümmte Stoßfedern des Birkhahns)
tail of hare 1., 2. tag for the end of fox tail	1. Blume (Hasenschwanz), 2. Fuchs-Luntenspitze
tail, breech, brush, drag	Fahne, Lunte, Rute, Standarte
tail, fan	Fächer (Auerhahnstoß)
tail, lyre shaped tail, playing tail (blackcock)	Leier, Schar, Schere, Spiel (Birkhahn)
tail, train (winged game)	Stoß (Federwild), Spiel
take (to) a battue	nehmen (Treiben nehmen)
take (to) care of	versorgen
take (to) food (wild boar)	Sauen nehmen Fraß auf
take (to) off the hunting dog from the track	abnehmen, abtragen
take (to) out the upper canine tooth (trophy) of red deer	auslösen der Grandeln (obere Eckzähne als Trophäe) beim Rotwild
take (to) prey by predatory birds	Beute machen durch Greifvögel
tallow (fat inside the body)	Talg (Fett aus dem Körper-inneren)
talon, claw	Fang (für Greifvögel), Gewaff, Waffen
target, buff patch, rump patch (hoofed game)	Spiegel (Schalenwild)
tarsus, leg, shank	Ständer

tassel, tuft of hairs	Quaste
tawny owl	Waldkauz
teal	Krickente
team of ducks, chain, covey of birds	Kette
tear-pit, lachrymal fossa	Tränengrube
teeth (canine tooth of red deer)	Grandeln, Haken (Cerviden)
teeth of comb	Balzfedern, Balzstifte, Zehenstifte
telescope	Spektiv
telescope sight	Zielfernrohr
telescope, field glass, binocular	Fernrohr, Fernglas, Jagdglas, Prismenglas
telling	Kugelschlag
Tengmalm´s owl	Raufußkauz
tercel	Terzel
terminal ballistics	Endballistik, Zielballistik
territorial boundary	Reviergrenze
territorial fight	Revierkampf
territory	Revier
test (to) the rifle or shotgun	Gewehr anschießen
testicle, coddes	Geschröte (Hoden von Hunden und Haarraubwild), Brunftkugel, Kurzwildbret
thanks to the hunters and beaters (drivers)	abdanken
thigh, clava, club, haunch, hind leg	Keule
thigh, flag	Hose (Federn an den Ständern mancher Greifvögel und Raufußhühner)
third verse of the courtship call	Hauptschlag (dritter Teil der Balzstrophe des Auerhahns)
three-barrelled gun	Drilling, Dreiläufer
throat, gula, gullet	Kehle
throat, gullet, oesophagus	Schlund

throat (neck of capercaillie and grouse)	Stingel (Hals von Auer-und Birkwild)
throat bot fly	Rachenbremse
throat feathers	Balzkragen
throw (to) (excrements of predatory birds)	schmeißen (Ausscheiden der Exkremente aller Greifvögel und Reiher)
throw (to) up (for all hoofed game, excluding wild boar)	aufwerfen
throw (to), to whelp	werfen, wölfen (gebären)
tibetan black bear	Kragenbär
tick	Zecke
tiger	Tiger
timber wolf	Timberwolf
time for giving birth, fawning season	Satzzeit, Setzzeit
time for pairing (bear)	Bärzeit
time of flight	Strichzeit
tine, branch, point, prong of antler	Ende, Geweihzacken, Sprosse
tine on the ground of the shovel antler, back antler	Dorn, Schauflerhaken
toe	Zehe
toils	Lappen
tom (male) cat	Kater
tongue	Graser, Lecker, Weidlöffel, Zunge
tooth	Zahn
tough, hard	hart
trace of stag in the middle of the lair, where the stag was sitting	Schlosstritt
trachea, windpipe	Drossel, Luftröhre
track (to), to circle	fährten, kreisen
track (to), to follow the trail of animals, to trail	abspüren, abfährten, spüren
track, footprint	Schalenabdruck, Trittsiegel

track, mark, sign, spoor	Pirschzeichen, Pürschzeichen
track, spoor, trail	Spur, Fährte, Wildspur
track (to) for wounded game	verlorensuchen
tracker	Spurensucher
tracker, trailer, boarhound	Finder, Saufinder, Packer
tracking, dogging, to search	absuchen, Suche, Suchjagd
tracks	Geläuf
trade in venison	Wildhandel
trail (to), to follow the trail of animals, to track	spüren, abfährten, abspüren
trail, track, spoor	Wildspur, Fährte, Spur
trail drag, scent bag, scented trail	Schleppe, Geschleife (künstliche Wundfährte für den Hund)
trailer, boarhound, tracker	Finder, Saufinder, Packer
trailing, searching with a hunting dog for wounded game, following a wounded game	Nachsuche
trail of a fleeing game	Fluchtfährte
train (to)	abführen, führen
train (to), to break in, to train the dog for hunting	abrichten, abführen, führen
train (winged game), tail	Stoß (Federwild), Spiel
trajectory	Flugbahn
trap	Eisen, Fangeisen
trap, snare	Falle
trapper	Fallensteller, Pelzjäger
trapping	Fallenjagd, Fangjagd
traverse (to)	durchziehen
tray	Mittelsprosse
tread (to) the hen	treten
tree of courtship or of mating	Balzbaum
tree on which hoofed game rubs	Malbaum
trespass right, the right to follow wounded game	Wildfolge, Nacheile
trichina	Trichine

trigger	Abzug
trigger guard	Abzugbügel
trigger safety	Abzugsicherung
troop, herd, pack	Rudel
trophy	Trophäe
trophy exposition	Trophäenschau
trophy stag	Erntehirsch
trot	Troll
trot (to) (fox, badger, wolf)	traben (Fuchs, Dachs, Wolf)
trunk, manubrium	Rüssel
truss the quarry, to shoot down upon prey	stoßen
tuft (heron, tufted duck, goosander)	Schopf
tuft, crest	Holle
tuft of hair, brush (hair at the end of pizzle of hoofed game)	Pinsel
tufted duck	Reiherente
tunnel, burrow	Bauröhren von Fuchs und Dachs (Geschleife)
turkey, turkey cock	Truthahn, Puter
turkey hen	Truthenne
turtle dove	Turteltaube
tusker, wild boar	Keiler
tusk of wild boar, weapon	Gewaff, Hauer, Waffen (Schwarzwild), auch Gewehre (untere Eckzähne des Keilers)
twelfe points stag	Zwölfender

U

udder, mammary glands of animals	Spinne, Gesäuge
uncock (to) the trigger	entstechen
uncouple (to), to let loose, to unleash	abhalsen, schnallen

under tailcoverts	Bruck (Unterschwanzdecken des Greifvogels)
uneven, odd	ungerade
unhuntsman-like	unweidmännisch
unlawful shooting, sacrilege against hunting	Jagdfrevel, Jagdvergehen
unleash (to), to uncouple; to let loose,	abhalsen, schnallen
unload (to)	entladen
unlock (to)	entsichern
unroll (to) the leash	abdocken (Schweißriemen)
upper breast, crop	Kropf (für Falken)
upper jaw, maxilla	Oberkiefer
upper tusk, grinder	Haderer
Ural owl	Habichtskauz
urinate (to), to piss	feuchten, nässen
urus, aurox, wildox	Auerochse
use (to) a gun	führen (Gewehr führen)
use (to) his nose for approaching the game (hunting dog)	anziehen

V

vaccination	Impfung
vagina (dogs, predaceous animals)	Schnalle
vagina, vulva of doe (hind)	Feuchtblatt
varios breeds, falcon	Jagdfalke
varvel	Drahle
veel, fawn, male calf of red deer	Hirschkalb
velvet (from animal horn or antler)	Bast, Gefege
velvet covered antlers	Bastgeweih
velvet mark	Fegestelle
venison	Wildbret

venison season	Feistzeit
ventilate (to), to air	lüften
venturi tube, pocket of gland	Saugloch, Schmalzröhre, Stinkloch
vermifuge	Wurmmittel
vermin, pest	Raubzeug
vesicle, bubble, blister, bladder	Blase
veterinary	Tierarzt
virus	Virus
vision, face, sight	Gesicht
vista, aisle	Waldschneise, Schneise
visual power	Sehschärfe
vixen, female of canids (fox), bitch fox, fuchsin	Bätze, Betze, Fähe, Füchsin
volatile, fleeing, on the run	flüchtig
vomits (of predatory birds), cast, pellet	Gewölle
vulture	Geier

W

wad	Pfropfen, Zwischenmittel
wag (to), to jog, to jolt	rütteln
walking up	Streife, Böhmische Streife
wall	Schalenrand, Sensel
wallow (to)	suhlen
wallow, wallowing place	Suhle
wapiti	Wapiti
warble fly	Dasselfliege
warren, enclosed pasture	Jagdgehege
warthog	Warzenschwein
waterfowl	Wasserwild
waterfowl hunting, wildfowl shooting	Wasserjagd
weak, of poor condition, little, small	gering, schwach

weak stag, young cock for capercaillie and black cock	Schneider
weapon, tusk of wild boar	Waffen, Gewaff, Hauer (Schwarzwild)
wear (to), a stag wears good or bad antlers	aufhaben
weasel	Mauswiesel, Wiesel
weather	Witterung (Wetter)
weather forecast	Wettervorhersage
webbed foot, palmate foot	Latschen, Ruder
weimarian dog	Weimaraner
western polecat	Iltis
Westphalian beagle	Westfälische Dachsbracke
whelp (to), to throw	werfen (gebären), wölfen
whelp, pup	Welpe
whet (to)	wetzen, schleifen
whine (to), anxious call	klagen
whistle (to)	pfeifen
whistle (to), to bleat, to call	blatten, fiepen
white back of certain mammals (ram), saddlecloth	Sattel, Schabracke, Schneefleck
white-fronted goose	Blässgans
whitetailed deer	Weißwedelhirsch
white-tailed eagle	Seeadler
whole bag of hooofed game, shooting	Abschuss, Gesamtabschuss
whooper swan	Singschwan
wig antlers	Perückengeweih
wigeon	Pfeifente
wild animal	Wildtier
wild boar	Schwarzwild, Wildschwein
wild boar, tusker	Keiler
wild cat	Wildkatze
wild duck	Wildente
wild fauna	Wildfauna, Wildtierbestand
wild goose	Wildgans
wild pig, sow	Sau
wild sow, cheek	Bache

wildebeest, gnu	Gnu
wilderness	Wildnis
wildfowl shooting, waterfowl hunting	Wasserjagd
wildlife management, game keeping, game protection	Hege, Wildhege, Wildschutz
wildlife research	Jagdwissenschaft
wildlife reserve, game reserve, wildlife park	Wildpark, Wildreservat, Wildschutzgebiet
wildox, aurox, urus	Auerochse
willow grouse	Moorschneehuhn
wind (to), to get wind off, to scent, to smell	winden, Wind bekommen, Wind holen, wittern, auch sichern, verhoffen
wind (to), to stop, to stand at gaze	sichern, verhoffen
wind, flatulence	Wind
windpipe, trachea	Luftröhre, Drossel
wing patch (capercaillie, black grouse)	Spiegel (Auer- und Birkwild)
wing, pinion	Flügel, Schwinge
wing (to)	flügeln
winged, broken-winged	geflügelt
winged game	Flugwild
wingshooting	Flugwildjagd
winter coat	Winterhaar
winter dormancy, winter restling,	Winterruhe
winter plumage, resting plumage	Winterkleid
witches path	Hexensteig
withers	Widerrist
wolf	Wolf
wolverine	Vielfraß
woodcock	Waldschnepfe
wood (great) grouse, capercaillie	Auerwild
wood lemming	Waldlemming

woodland hare	Waldhase
wood pigeon	Ringeltaube
wool, hair	Wolle
work	Arbeit
worm (to), to search for food (snipe), to pierce, to sting	stechen, wurmen (Schnepfe)
wound (to) a game, to injure	anschießen, anschweißen, krankschießen
wounded in the bowels	weidwund
wrist	Kolbenhals

Y

yak	Yak
yearling buck (second year)	Jährling
yelp (to), give tongue, to bark, also by foxes during the rutting season	Lautgeben des Jagdhundes, anschlagen, bellen, verbellen, auch Lautäußerung der Füchse während der Ranz
young animal, juvenile	Junge
young game	Jungwild
young goshawk	Rothabicht
young seal	Heuler
young tusker	Überläuferkeiler
young wild boar, gruntling	Frischling

Z

zebra	Zebra
zebu	Zebu
zero (to)	einschießen
zoologica garden, zoo	Tiergarten
zoology	Tierkunde
zoonosis, pest	Tierseuche

Literaturverzeichnis

Edmund Launert	Biologisches Wörterbuch – Deutsch-Englisch • Englisch-Deutsch. Verlag Eugen Ulmer, Stuttgart. 1998.
Jürgen Schulte	in Biologisches Wörterbuch – Deutsch-Englisch • Englisch-Deutsch. Verlag Eugen Ulmer, Stuttgart, 1998.
	Die Jägersprache – kurz gefasst. Verlag J. Neumann-Neudamm, Melsungen, 3. Auflage 2021.
Attila Zoltan	Bildwörterbuch der Jagd, Nimrod, Vadaszuisag.
Joachim Köller Istvan Miko Antal Török ÖdönZoltan	Denes Natur Mühely, 2000.

Quellen

Attila Zoltan	**Waffenkundliche Begriffe** in Bildwörterbuch der Jagd. Nimrod, Vadaszuisag. Denes Natur Mühely, 2000.
EdmundLaunert	**Englische Namensgebung für alle nicht heimischen Tierarten** in Biologisches Wörterbuch – Deutsch-Englisch • Englisch-Deutsch. Verlag Eugen Ulmer, Stuttgart. 1998.
Ulrich Sedlag	**Veterinärmedizin (Anatomie der inneren Organe, Wildkrankheiten)** in Biologisches Wörterbuch – Deutsch-Englisch • Englisch-Deutsch. Verlag Eugen Ulmer, Stuttgart. 1998.

ATTENTION
reversible
brochure!